Springer Topics in Signal Processing

Volume 18

More information about this series at http://www.springer.com/series/8109

Jacob Benesty · Israel Cohen ·
Jingdong Chen

Array Processing

Kronecker Product Beamforming

Springer

Jacob Benesty
INRS-EMT
University of Quebec
Montreal, QC, Canada

Israel Cohen
Department of Electrical Engineering
Technion—Israel Institute of Technology
Haifa, Israel

Jingdong Chen
School of Marine Science and Technology
Northwestern Polytechnical University
Xi'an, China

ISSN 1866-2609 ISSN 1866-2617 (electronic)
Springer Topics in Signal Processing
ISBN 978-3-030-15599-5 ISBN 978-3-030-15600-8 (eBook)
https://doi.org/10.1007/978-3-030-15600-8

Library of Congress Control Number: 2019934360

This Springer imprint is published by the registered company Springer Nature Switzerland AG
The registered company address is: Gewerbestrasse 11, 6330 Cham, Switzerland

Abstract

The focus of this book is on array processing and beamforming with Kronecker products. We consider a large family of sensor arrays that enable the decomposition of the steering vector as a Kronecker product of two steering vectors of smaller virtual arrays. Instead of directly designing a global beamformer for the original array, we break it down following the decomposition of the steering vector, and design smaller virtual beamformers separately optimized for each virtual array. This implies smaller matrices to invert, which increases the robustness of the beamformers, and less observations to estimate the statistics when necessary. We explain how to perform beamforming with Kronecker product filters differently from the well-known and studied conventional approach. We show how to derive fixed, adaptive, and differential beamformers with remarkable flexibility thanks to the Kronecker product formulation. Furthermore, fixed and adaptive beamformers can be combined very intelligently, so that the best of each one of these two approaches is emphasized for performance enhancement. We also address the problem of spatiotemporal signal enhancement, and explain how to perform Kronecker product filtering in this context.

Contents

Acronyms

2-D	two dimensional
C	cardioid
cFA	combined fixed/adaptive
D	dipole
DF	directivity factor
DS	delay and sum
FBR	front-to-back ratio
H	hypercardioid
LCMV	linearly constrained minimum variance
mDF	maximum directivity factor
MN	minimum norm
MSE	mean-squared error
MVDR	minimum variance distortionless response
NS	null steering
NULA	nonuniform linear array
PmDF	partial maximum directivity factor
PS	partial superdirective
RA	rectangular array
S	supercardioid
SD	superdirective
SNR	signal-to-noise ratio
STFT	short-time Fourier transform
ULA	uniform linear array
W	Wiener
WNG	white noise gain

Chapter 1
Introduction

In this chapter, we explain what is beamforming, how it works, and introduce beamforming with Kronecker products. Then, we present the organization of this work.

1.1 Beamforming

Beamforming or spatial filtering is an active and central research area of array signal processing [1, 2, 3, 4, 5]. Beamforming algorithms fall into two major categories depending on whether the noise or signal statistics are considered in forming the beamforming filters, i.e., fixed and adaptive beamforming. A fixed beamformer is a spatial filter that has the ability to form a main beam in the direction of the desired signal and, possibly, place nulls in the directions of interferences without the knowledge of the data picked up by the array or the statistics of the desired and noise signals. Accordingly, the coefficients of this filter are fixed and do not depend on the changes of the wave propagation environment in which the array performs. Adaptive beamforming algorithms consider using either the noise statistics or the statistics of the array observation data to optimize the beamforming filters. The performance of adaptive beamforming can be "more" optimal than its fixed counterpart as long as the signal statistics are correctly estimated.

Consider an array of M sensors and assume that a desired signal impinges on this array from the angle θ_{d}, as illustrated in Fig. 1.1. Then, the observation signal vector (of length M) is

$$\begin{aligned} \mathbf{y}(\omega) &= \begin{bmatrix} Y_1(\omega) & Y_2(\omega) & \cdots & Y_M(\omega) \end{bmatrix}^T \\ &= \mathbf{d}_{\theta_{\mathrm{d}}}(\omega) X(\omega) + \mathbf{v}(\omega), \end{aligned} \tag{1.1}$$

J. Benesty et al., *Array Processing*, Springer Topics in Signal Processing 18, https://doi.org/10.1007/978-3-030-15600-8_1

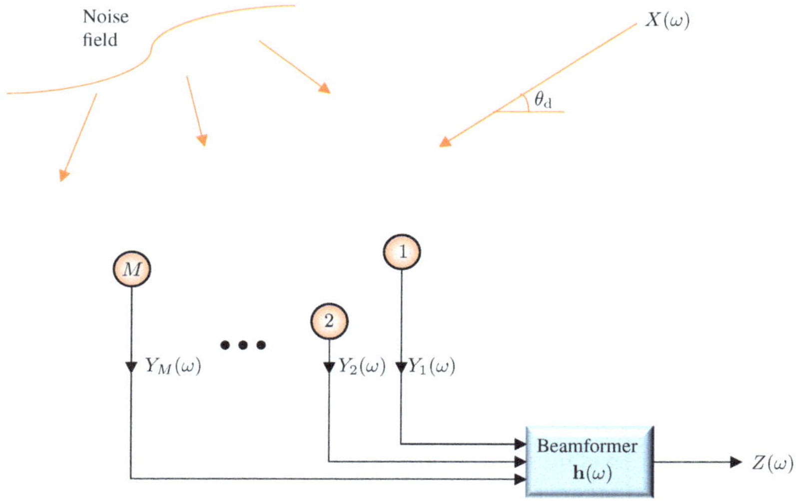

Fig. 1.1 Illustration of an array of M sensors and a beamformer applied to the observation signals.

where ω is the angular frequency, $Y_m(\omega)$ is the mth sensor signal, $X(\omega)$ is the desired source signal, $\mathbf{d}_{\theta_d}(\omega)$ is the steering vector at $\theta = \theta_d$, and $\mathbf{v}(\omega)$ is the additive noise signal vector defined similarly to $\mathbf{y}(\omega)$.

The conventional way of doing beamforming is by applying a complex-valued linear filter, $\mathbf{h}(\omega)$, of length M to the observation signal vector, $\mathbf{y}(\omega)$ [6, 7]. This gives the output:

$$Z(\omega) = \mathbf{h}^H(\omega)\,\mathbf{y}(\omega), \tag{1.2}$$

where $Z(\omega)$ is the estimate of the desired signal, $X(\omega)$. The process of finding the appropriate filter, $\mathbf{h}(\omega)$, based on some given performance criteria is called beamforming.

One of the most useful performance criteria is the beampattern, which describes the sensitivity of the beamformer to a plane wave impinging on the array from the direction θ. Mathematically, it is defined as

$$\mathcal{B}_\theta[\mathbf{h}(\omega)] = \mathbf{d}_\theta^H(\omega)\,\mathbf{h}(\omega). \tag{1.3}$$

A convenient way to evaluate the sensitivity of the array to some of its imperfections, such as sensor noise, is via the so-called white noise gain (WNG), which is given by

$$\mathcal{W}[\mathbf{h}(\omega)] = \frac{\left|\mathbf{h}^H(\omega)\,\mathbf{d}_{\theta_d}(\omega)\right|^2}{\mathbf{h}^H(\omega)\,\mathbf{h}(\omega)}. \tag{1.4}$$

Another important measure, which quantifies how the sensor array performs in the presence of reverberation is the directivity factor (DF), defined as (for linear arrays) [4]

$$\mathcal{D}\left[\mathbf{h}(\omega)\right] = \frac{\left|\mathcal{B}_{\theta_{\mathrm{d}}}\left[\mathbf{h}(\omega)\right]\right|^2}{\frac{1}{2}\int_0^{\pi}\left|\mathcal{B}_{\theta}\left[\mathbf{h}(\omega)\right]\right|^2 \sin\theta d\theta} = \frac{\left|\mathbf{h}^H(\omega)\,\mathbf{d}_{\theta_{\mathrm{d}}}(\omega)\right|^2}{\mathbf{h}^H(\omega)\,\boldsymbol{\Gamma}(\omega)\,\mathbf{h}(\omega)}, \tag{1.5}$$

where

$$\boldsymbol{\Gamma}(\omega) = \frac{1}{2}\int_0^{\pi} \mathbf{d}_{\theta}(\omega)\,\mathbf{d}_{\theta}^H(\omega)\sin\theta d\theta. \tag{1.6}$$

With the conventional beamforming approach, M coefficients of $\mathbf{h}(\omega)$ need to be estimated. Here, we are interested in arrays of $M = M_1 M_2$ sensors that enable to decompose the steering vector as a Kronecker product of two steering vectors of smaller virtual arrays [8, 9, 10]. That is, the steering vector (of length M) is decomposed as

$$\mathbf{d}_{\theta}(\omega) = \mathbf{d}_{1,\theta}(\omega) \otimes \mathbf{d}_{2,\theta}(\omega), \tag{1.7}$$

where $\otimes$ is the Kronecker product, $\mathbf{d}_{1,\theta}(\omega)$ is the steering vector (of length M_1) corresponding to a virtual array of M_1 sensors, and $\mathbf{d}_{2,\theta}(\omega)$ is the steering vector (of length M_2) corresponding to another virtual array of M_2 sensors.

Figure 1.2 shows examples of Kronecker product decompositions of a uniform linear array (ULA) of 12 sensors with an interelement spacing equal to δ into two smaller virtual ULAs with different numbers of sensors and different interelement spacings. Note that (1.7) is satisfied whenever the global array can be obtained from replications of one virtual array to the sensor positions of the other virtual array. Figure 1.3 shows examples of Kronecker product decompositions of a nonuniform linear array (NULA) of 12 sensors into two smaller virtual ULAs. Figure 1.4 shows examples of Kronecker product decompositions of a two-dimensional rectangular array (RA) of 24 sensors into ULAs and RAs with different numbers of sensors and different interelement spacings.

In the proposed approach, instead of directly designing the filter $\mathbf{h}(\omega)$ of length M, we break it down following the decomposition of the global steering vector as

$$\mathbf{h}(\omega) = \mathbf{h}_1(\omega) \otimes \mathbf{h}_2(\omega), \tag{1.8}$$

where $\mathbf{h}_1(\omega)$ and $\mathbf{h}_2(\omega)$ are two complex-valued linear filters of length M_1 and M_2, respectively. With this method, we only need to estimate $M_1 + M_2$

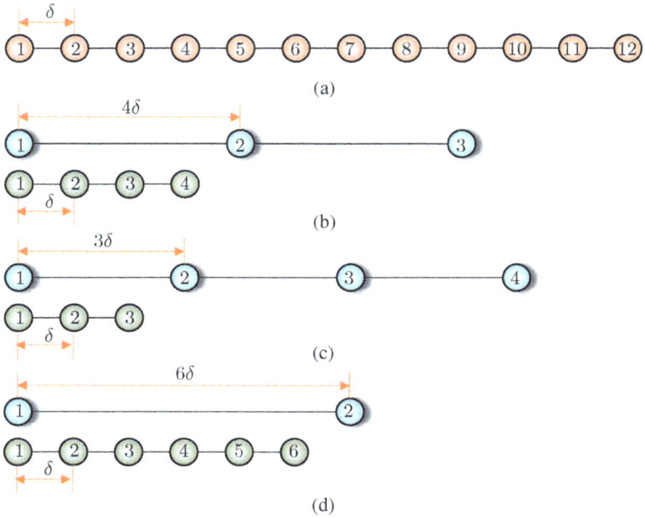

Fig. 1.2 Examples of Kronecker product decompositions of a ULA into two smaller virtual ULAs: (a) a ULA of 12 sensors with an interelement spacing equal to δ; (b) one virtual ULA of 3 sensors with an interelement spacing equal to 4δ, and another virtual ULA of 4 sensors with an interelement spacing equal to δ; (c) one virtual ULA of 4 sensors with an interelement spacing equal to 3δ, and another virtual ULA of 3 sensors with an interelement spacing equal to δ; and (d) one virtual ULA of 2 sensors with an interelement spacing equal to 6δ, and another virtual ULA of 6 sensors with an interelement spacing equal to δ.

coefficients [M_1 for $\mathbf{h}_1(\omega)$ and M_2 for $\mathbf{h}_2(\omega)$] instead of $M = M_1 M_2$ for the conventional technique. This implies smaller matrices to invert (increasing robustness) and less observations to estimate the statistics when necessary.

There are many ways to optimize the coefficients of $\mathbf{h}_1(\omega)$ and $\mathbf{h}_2(\omega)$ depending on what we want and the application at hand. We introduce fixed and adaptive Kronecker product beamformers, and show how to derive such beamformers, as well as new approaches through the design of $\mathbf{h}_1(\omega)$ and $\mathbf{h}_2(\omega)$. We also show how to combine very intelligently fixed and adaptive beamformers, so that the best of each one of these two approaches is emphasized for performance enhancement.

The decomposition (1.8) enables to perform beamforming differently from the well-known and studied conventional approach. In our context, the global beampattern can be expressed as the product of two beamformer beampatterns:

$$\mathcal{B}_\theta\left[\mathbf{h}(\omega)\right] = \mathcal{B}_{1,\theta}\left[\mathbf{h}_1(\omega)\right] \times \mathcal{B}_{2,\theta}\left[\mathbf{h}_2(\omega)\right], \tag{1.9}$$

where

$$\mathcal{B}_{1,\theta}\left[\mathbf{h}_1(\omega)\right] = \mathbf{d}_{1,\theta}^H(\omega)\,\mathbf{h}_1(\omega) \tag{1.10}$$

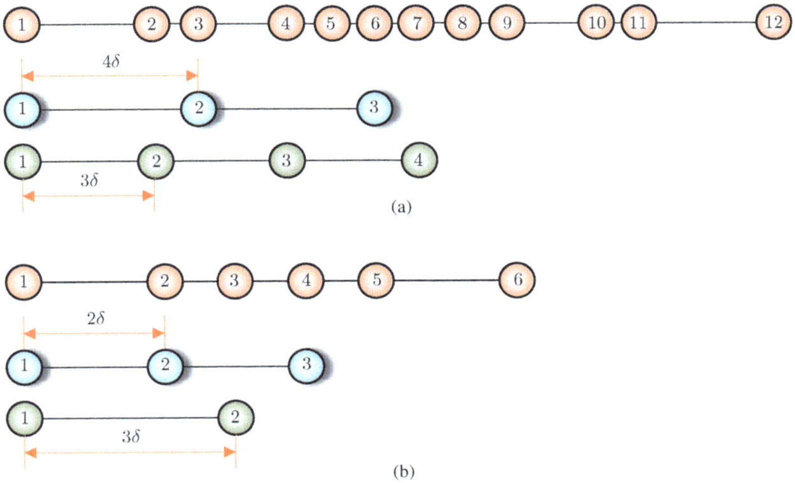

Fig. 1.3 Examples of Kronecker product decompositions of a NULA into two smaller virtual ULAs: (a) a nonuniform linear array of 12 sensors, one virtual ULA of 3 sensors with an interelement spacing equal to 4δ, and another virtual ULA of 4 sensors with an interelement spacing equal to 3δ; and (b) a nonuniform linear array of 6 sensors, one virtual ULA of 3 sensors with an interelement spacing equal to 2δ, and another virtual ULA of 2 sensors with an interelement spacing equal to 3δ.

is the beampattern of the first virtual array, and

$$\mathcal{B}_{2,\theta}\left[\mathbf{h}_2(\omega)\right] = \mathbf{d}_{2,\theta}^H(\omega)\,\mathbf{h}_2(\omega) \tag{1.11}$$

is the beampattern of the second virtual array. Also, the WNG of the global array can be expressed as the product of the WNGs of the first and second virtual arrays, while the DF of the global array cannot be factorized as the product of the DFs of the two virtual ULAs, i.e.,

$$\mathcal{W}\left[\mathbf{h}(\omega)\right] = \mathcal{W}_1\left[\mathbf{h}_1(\omega)\right] \times \mathcal{W}_2\left[\mathbf{h}_2(\omega)\right], \tag{1.12}$$

$$\mathcal{D}\left[\mathbf{h}(\omega)\right] \neq \mathcal{D}_1\left[\mathbf{h}_1(\omega)\right] \times \mathcal{D}_2\left[\mathbf{h}_2(\omega)\right]. \tag{1.13}$$

These interesting properties can be exploited in the design of very flexible global beamformers.

1.2 Organization of the Work

The material in this book is organized into seven chapters, including this one.

In Chapter 2, we formulate the problem of Kronecker product beamforming with ULAs. We define some important performance measures in this

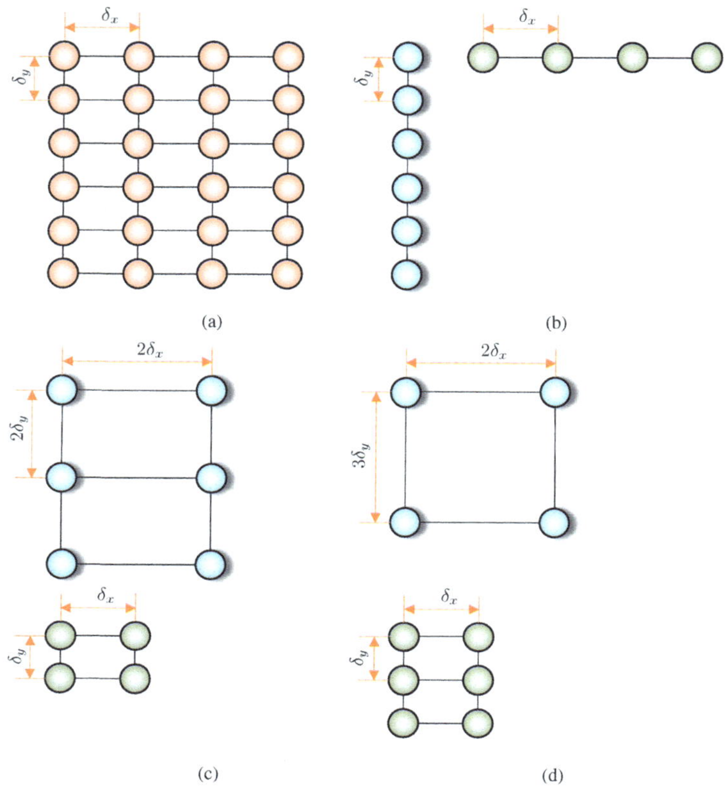

Fig. 1.4 Examples of Kronecker product decompositions of an RA into two smaller virtual arrays: (a) an RA of 24 sensors with an interelement spacing equal to δ_x along the x axis and an interelement spacing equal to δ_y along the y axis; (b) one virtual ULA of 6 sensors with an interelement spacing equal to δ_y and another virtual ULA of 4 sensors with an interelement spacing equal to δ_x; (c) one virtual RA of 2×3 sensors with an interelement spacing equal to $2\delta_x$ along the x axis and an interelement spacing equal to $2\delta_y$ along the y axis, and another virtual RA of 2×2 sensors with an interelement spacing equal to δ_x along the x axis and an interelement spacing equal to δ_y along the y axis; and (d) one virtual RA of 2×2 sensors with an interelement spacing equal to $2\delta_x$ along the x axis and an interelement spacing equal to $3\delta_y$ along the y axis, and another virtual RA of 2×3 sensors with an interelement spacing equal to δ_x along the x axis and an interelement spacing equal to δ_y along the y axis.

context, and explain how to perform beamforming with Kronecker product filters differently from the well-known and studied conventional approach.

In Chapter 3, we show how to derive fixed, adaptive, and differential beamformers with remarkable flexibility thanks to the Kronecker product formulation. We also introduce new beamforming approaches that combine fixed and adaptive beamformers, so that the best of each one of these two approaches is emphasized for performance enhancement.

In Chapter 4, we generalize the Kronecker product beamforming with other decompositions of the steering vector associated with ULAs. We show how to design important differential beamformers, in a very elegant way, thanks to the Kronecker product decompositions of the steering vector and the proposed filtering approach.

The focus of Chapter 5 is on NULAs. We show how from two virtual ULAs we can construct a physical NULA whose associated steering vector is the Kronecker product of the steering vectors associated with the virtual arrays. Then, we explain how Kronecker product beamforming is performed and derive some interesting optimal beamformers.

Chapter 6 continues the investigation of Kronecker product beamforming, but with two-dimensional arrays. Conventional two-dimensional beamforming methods suffer from the need to invert very ill-conditioned large matrices, which necessarily lead to serious estimation problems in the presence of uncertainties. We show how to avoid this problem with Kronecker product beamforming, and how to extend some of the results obtained in previous chapters to two-dimensional arrays such as the rectangular ones.

Finally, in Chapter 7, we address the problem of spatiotemporal signal enhancement with any array geometry. We show how the Kronecker product appears naturally in the definition of the signal vector by taking into account the interframe correlation. We derive spatiotemporal Kronecker product filters, and explain how to perform noise reduction with the most well-known performance measures.

References

1. J. Benesty, J. Chen, and Y. Huang, *Microphone Array Signal Processing.* Berlin, Germany: Springer-Verlag, 2008.
2. J. Benesty, I. Cohen, and J. Chen, *Fundamentals of Signal Enhancement and Array Signal Processing.* Singapore: Wiley–IEEE Press, 2018.
3. S. Haykin and K. J. R. Liu, *Handbook on Array Processing and Sensor Networks.* Hoboken, NJ: Wiley & Sons, 2010.
4. H. L. Van Trees, *Detection, Estimation, and Modulation Theory, Optimum Array Processing* (Part IV). Hoboken, NJ: Wiley-Interscience, 2002.
5. D. H. Johnson and D. E. Dudgeon, *Array Signal Processing: Concepts and Techniques.* Upper Saddle River, NJ: Prentice Hall, 1993.
6. S. Haykin, *Array Signal Processing.* Upper Saddle River, NJ: Prentice-Hall, 1984.
7. H. Krim and M. Viberg, "Two decades of array signal processing research: the parametric approach," *IEEE Signal Process. Mag.*, vol. 13, pp.67–94, Jul. 1996.
8. Y. I. Abramovich, G. J. Frazer, and B. A. Johnson, "Iterative adaptive Kronecker MIMO radar beamformer: description and convergence analysis," *IEEE Trans. Signal Process.*, vol. 58, pp. 3681–3691, Jul. 2010.
9. F. P. Ribeiro and V. H. Nascimento, "Fast transforms for acoustic imaging–Part I: Theory," *IEEE Trans. Image Process.*, vol. 20, pp. 2229–2240, Aug. 2011.
10. B. Masiero and V. H. Nascimento, "Revisiting the Kronecker array transform," *IEEE Signal Process. Lett.*, vol. 24, pp. 525–529, May 2017.

Chapter 2
Problem Formulation with Uniform Linear Arrays

In this chapter, we describe ULAs and define the associated steering vector. In one very particular case, we explain how this vector can be decomposed as a Kronecker product of two steering vectors of smaller ULAs with the same number of elements. Thanks to this decomposition, we explain how to perform beamforming with Kronecker product filters. Then, we derive some very important and useful performance measures in this context that will be of great help to design and evaluate all kind of beamformers.

2.1 Signal Model

We consider a ULA consisting of $M = M_0^2$ omnidirectional microphones, where $M_0 \geq 2$, with an interelement spacing equal to δ. Typical and practical values of M are 4, 9, 16, and even 25. We will refer to this array as the global or whole ULA. Now, assume that a farfield desired source signal (plane wave) propagates from the azimuth angle, θ, in an anechoic acoustic environment at the speed of sound, i.e., $c = 340$ m/s, and impinges on the above described microphone array (see Fig. 2.1). Then, the corresponding steering vector (of length M) is [1], [2]

$$\mathbf{d}_\theta(\omega) = \begin{bmatrix} 1 & e^{-\jmath\varpi(\theta)} & e^{-\jmath 2\varpi(\theta)} & \cdots & e^{-\jmath(M-1)\varpi(\theta)} \end{bmatrix}^T, \quad (2.1)$$

where the superscript ^{T} is the transpose operator, $\jmath$ is the imaginary unit,

$$\varpi(\theta) = \frac{\omega\delta\cos\theta}{c}, \quad (2.2)$$

$\omega = 2\pi f$ is the angular frequency, and $f > 0$ is the temporal frequency. Since $\cos\theta$ is an even function so is $\mathbf{d}_\theta(\omega)$. Therefore, the study with linear arrays is limited to angles $\theta \in [0, \pi]$.

J. Benesty et al., *Array Processing*, Springer Topics in Signal Processing 18, https://doi.org/10.1007/978-3-030-15600-8_2

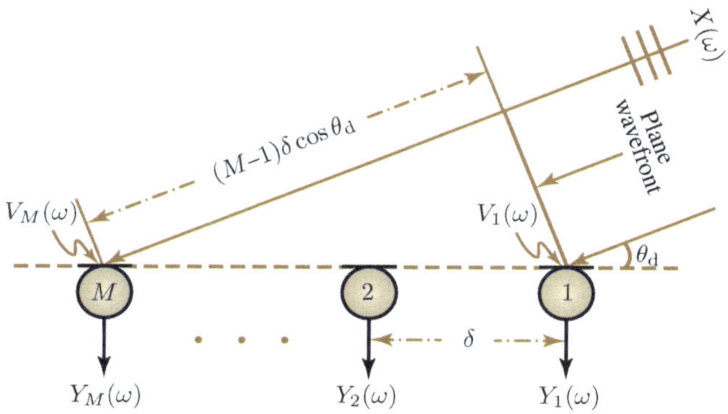

Fig. 2.1 A uniform linear array with M sensors.

The very interesting Vandermonde structure of the vector $\mathbf{d}_\theta(\omega)$ can be exploited. Indeed, it is easy to see that this steering vector can be decomposed as

$$\mathbf{d}_\theta(\omega) = \mathbf{d}_{1,\theta}(\omega) \otimes \mathbf{d}_{2,\theta}(\omega), \tag{2.3}$$

where $\otimes$ is the Kronecker product,

$$\mathbf{d}_{1,\theta}(\omega) = \begin{bmatrix} 1 & e^{-\jmath M_0 \varpi(\theta)} & e^{-\jmath 2 M_0 \varpi(\theta)} & \cdots & e^{-\jmath M_0 (M_0-1)\varpi(\theta)} \end{bmatrix}^T \tag{2.4}$$

is the steering vector (of length M_0) corresponding to a ULA of M_0 sensors with an interelement spacing equal to $M_0\delta$, and

$$\mathbf{d}_{2,\theta}(\omega) = \begin{bmatrix} 1 & e^{-\jmath \varpi(\theta)} & e^{-\jmath 2 \varpi(\theta)} & \cdots & e^{-\jmath (M_0-1)\varpi(\theta)} \end{bmatrix}^T \tag{2.5}$$

is the steering vector (of length M_0) corresponding also to a ULA of M_0 sensors but with an interelement spacing equal to δ. In fact, the components of $\mathbf{d}_{2,\theta}(\omega)$ are identical to the first M_0 elements of $\mathbf{d}_\theta(\omega)$. We will refer to these two arrays as the first and second ULAs or the subarrays. In the design of beamformers or beampatterns, it is important to keep in mind that the second ULA is much less sensitive to spatial aliasing than the first one since $\delta \ll M_0\delta$. To summarize, we can state that, when $M = M_0^2$, the steering vector (of length M_0^2) associated with the global ULA is simply the Kronecker product of the steering vectors (of length M_0) associated with the first and second ULAs.

Assume that the desired signal propagates from the angle θ_d. Then, the observation signal vector (of length M) is [3], [4]

$$\begin{aligned}\mathbf{y}(\omega) &= \begin{bmatrix} Y_1(\omega) & Y_2(\omega) & \cdots & Y_M(\omega) \end{bmatrix}^T \\ &= \mathbf{x}(\omega) + \mathbf{v}(\omega) \\ &= \mathbf{d}_{\theta_\mathrm{d}}(\omega) X(\omega) + \mathbf{v}(\omega), \end{aligned} \tag{2.6}$$

where $Y_m(\omega)$ is the mth microphone signal, $\mathbf{x}(\omega) = \mathbf{d}_{\theta_\mathrm{d}}(\omega) X(\omega)$, $X(\omega)$ is the zero-mean desired source signal, $\mathbf{d}_{\theta_\mathrm{d}}(\omega) = \mathbf{d}_{1,\theta_\mathrm{d}}(\omega) \otimes \mathbf{d}_{2,\theta_\mathrm{d}}(\omega)$ is the steering vector at $\theta = \theta_\mathrm{d}$ (direction of the desired source), $\mathbf{v}(\omega)$ is the zero-mean additive noise signal vector defined similarly to $\mathbf{y}(\omega)$, and $X(\omega)$ and $\mathbf{v}(\omega)$ are uncorrelated. In the rest, in order to simplify the notation, we drop the dependence on the angular frequency, ω. So, for example, (2.6) is written as $\mathbf{y} = \mathbf{d}_{\theta_\mathrm{d}} X + \mathbf{v}$. We deduce that the covariance matrix of $\mathbf{y}$ is

$$\begin{aligned}\mathbf{\Phi}_\mathbf{y} &= E\left(\mathbf{y}\mathbf{y}^H\right) \\ &= \mathbf{\Phi}_\mathbf{x} + \mathbf{\Phi}_\mathbf{v}, \end{aligned} \tag{2.7}$$

where $E(\cdot)$ denotes mathematical expectation, the superscript ^{H} is the conjugate-transpose operator,

$$\begin{aligned}\mathbf{\Phi}_\mathbf{x} &= \phi_X \mathbf{d}_{\theta_\mathrm{d}} \mathbf{d}_{\theta_\mathrm{d}}^H \\ &= \phi_X \left(\mathbf{d}_{1,\theta_\mathrm{d}} \otimes \mathbf{d}_{2,\theta_\mathrm{d}}\right) \left(\mathbf{d}_{1,\theta_\mathrm{d}} \otimes \mathbf{d}_{2,\theta_\mathrm{d}}\right)^H \\ &= \phi_X \left(\mathbf{d}_{1,\theta_\mathrm{d}} \mathbf{d}_{1,\theta_\mathrm{d}}^H\right) \otimes \left(\mathbf{d}_{2,\theta_\mathrm{d}} \mathbf{d}_{2,\theta_\mathrm{d}}^H\right) \end{aligned} \tag{2.8}$$

is the covariance matrix of $\mathbf{x}$, with $\phi_X = E\left(|X|^2\right)$ being the variance of X, and $\mathbf{\Phi}_\mathbf{v} = E\left(\mathbf{v}\mathbf{v}^H\right)$ is the covariance matrix of $\mathbf{v}$. Assuming that the first sensor is the reference, we can express (2.7) as

$$\mathbf{\Phi}_\mathbf{y} = \phi_X \mathbf{d}_{\theta_\mathrm{d}} \mathbf{d}_{\theta_\mathrm{d}}^H + \phi_{V_1} \mathbf{\Gamma}_\mathbf{v}, \tag{2.9}$$

where $\phi_{V_1} = E\left(|V_1|^2\right)$ is the variance of the noise at the reference sensor and $\mathbf{\Gamma}_\mathbf{v} = \mathbf{\Phi}_\mathbf{v}/\phi_{V_1}$ is the pseudo-coherence matrix of the noise. In the case of the spherically isotropic (diffuse) noise field, which will be assumed in fixed beamforming, (2.9) becomes

$$\mathbf{\Phi}_\mathbf{y} = \phi_X \mathbf{d}_{\theta_\mathrm{d}} \mathbf{d}_{\theta_\mathrm{d}}^H + \phi \mathbf{\Gamma}, \tag{2.10}$$

where ϕ is the variance of the diffuse noise and

$$\mathbf{\Gamma} = \frac{1}{2} \int_0^\pi \mathbf{d}_\theta \mathbf{d}_\theta^H \sin\theta d\theta. \tag{2.11}$$

It can be verified that the elements of the $M \times M$ matrix $\mathbf{\Gamma}(\omega)$ are

$$\begin{aligned} [\boldsymbol{\Gamma}(\omega)]_{ij} &= \frac{\sin[\omega(j-i)\delta/c]}{\omega(j-i)\delta/c} \\ &= \mathrm{sinc}[\omega(j-i)\delta/c], \end{aligned} \tag{2.12}$$

with $[\boldsymbol{\Gamma}(\omega)]_{mm} = 1,\ m = 1, 2, \ldots, M$.

One of our main objectives in this study is to take advantage of the global ULA steering vector structure in the particular case of $M = M_0^2$ to perform beamforming differently from the well-known and studied conventional approach. It will be demonstrated that the new technique is extremely flexible.

2.2 Beamforming with Kronecker Product Filters

The conventional way of doing beamforming is by applying a complex-valued linear filter, $\mathbf{h}_\mathrm{C}$, of length M to the observation signal vector, $\mathbf{y}$. This processing is [4]

$$Z_\mathrm{C} = \mathbf{h}_\mathrm{C}^H \mathbf{y}, \tag{2.13}$$

where Z_C is the estimate of the desired signal, X. While this approach is optimal as far as linear filtering is concerned, it lacks flexibility and M_0^2 coefficients of $\mathbf{h}_\mathrm{C}$ need to be estimated.

In order to fully exploit the structure of the global steering vector, let us consider the $M \times 1$ complex-valued filters having the form:

$$\mathbf{h} = \mathbf{h}_1 \otimes \mathbf{h}_2, \tag{2.14}$$

where $\mathbf{h}_1$ and $\mathbf{h}_2$ are two complex-valued linear filters of length M_0. In other words, the global beamformer, $\mathbf{h}$, follows the decomposition of the global steering vector, $\mathbf{d}_\theta$. In the proposed approach, beamforming is performed by applying $\mathbf{h}$ [as defined in (2.14)] to $\mathbf{y}$ [from (2.6)]. We get

$$\begin{aligned} Z &= \mathbf{h}^H \mathbf{y} \\ &= \mathbf{h}^H \mathbf{d}_{\theta_\mathrm{d}} X + \mathbf{h}^H \mathbf{v} \\ &= X_\mathrm{fd} + V_\mathrm{rn}, \end{aligned} \tag{2.15}$$

where Z is the estimate of the desired signal, X,

$$\begin{aligned} X_\mathrm{fd} &= (\mathbf{h}_1 \otimes \mathbf{h}_2)^H (\mathbf{d}_{1,\theta_\mathrm{d}} \otimes \mathbf{d}_{2,\theta_\mathrm{d}}) X \\ &= \left(\mathbf{h}_1^H \mathbf{d}_{1,\theta_\mathrm{d}}\right) \left(\mathbf{h}_2^H \mathbf{d}_{2,\theta_\mathrm{d}}\right) X \end{aligned} \tag{2.16}$$

is the filtered desired signal, and

$$V_\mathrm{rn} = (\mathbf{h}_1 \otimes \mathbf{h}_2)^H \mathbf{v} \tag{2.17}$$

is the residual noise. We deduce that the variance of Z is

$$\phi_Z = \phi_X \left| \mathbf{h}_1^H \mathbf{d}_{1,\theta_\mathrm{d}} \right|^2 \left| \mathbf{h}_2^H \mathbf{d}_{2,\theta_\mathrm{d}} \right|^2 + \phi_{V_1} \left(\mathbf{h}_1 \otimes \mathbf{h}_2 \right)^H \mathbf{\Gamma}_\mathbf{v} \left(\mathbf{h}_1 \otimes \mathbf{h}_2 \right). \quad (2.18)$$

We see that with this method, we only need to estimate $2M_0$ coefficients (M_0 for $\mathbf{h}_1$ and M_0 for $\mathbf{h}_2$) instead of M_0^2 for the conventional technique. This implies smaller matrices to invert (increasing robustness) and less observations to estimate the statistics when necessary. Notice that this way of doing beamforming may not be completely new. A similar approach was proposed in [5] but in the context of MIMO radar applications and in a rather very limited way.

In our context, the distortionless constraint in the direction of the desired source, i.e., $\theta = \theta_\mathrm{d}$, is often required, i.e.,

$$\mathbf{h}^H \mathbf{d}_{\theta_\mathrm{d}} = \left(\mathbf{h}_1^H \mathbf{d}_{1,\theta_\mathrm{d}} \right) \left(\mathbf{h}_2^H \mathbf{d}_{2,\theta_\mathrm{d}} \right) = 1. \quad (2.19)$$

Therefore, we will often (if not always) choose $\mathbf{h}_1^H \mathbf{d}_{1,\theta_\mathrm{d}} = \mathbf{h}_2^H \mathbf{d}_{2,\theta_\mathrm{d}} = 1$, so that (2.19) is satisfied.

2.3 Performance Measures

We are going to define some important performance measures by using Kronecker product filters. It will be shown how flexibility appears thanks to this decomposition.

The first useful measure discussed in this section is the beampattern or directivity pattern, which describes the sensitivity of the beamformer to a plane wave (source signal) impinging on the global ULA from the direction θ. Mathematically, it is defined as

$$\begin{aligned} \mathcal{B}_\theta (\mathbf{h}) &= \mathbf{d}_\theta^H \mathbf{h} \quad (2.20) \\ &= \left(\mathbf{d}_{1,\theta}^H \mathbf{h}_1 \right) \left(\mathbf{d}_{2,\theta}^H \mathbf{h}_2 \right) \\ &= \mathcal{B}_{1,\theta} (\mathbf{h}_1) \times \mathcal{B}_{2,\theta} (\mathbf{h}_2), \end{aligned}$$

where

$$\begin{aligned} \mathcal{B}_{1,\theta} (\mathbf{h}_1) &= \mathbf{d}_{1,\theta}^H \mathbf{h}_1 \\ &= \sum_{m=1}^{M_0} H_{1,m} e^{\jmath (m-1) M_0 \varpi(\theta)} \quad (2.21) \end{aligned}$$

is the beampattern of the first ULA, with $H_{1,m}$, $m = 1, 2, \ldots, M_0$ being the coefficients of $\mathbf{h}_1$, and

$$\begin{aligned}\mathcal{B}_{2,\theta}\left(\mathbf{h}_2\right) &= \mathbf{d}_{2,\theta}^H \mathbf{h}_2 \\ &= \sum_{m=1}^{M_0} H_{2,m} e^{\jmath(m-1)\varpi(\theta)} \end{aligned} \tag{2.22}$$

is the beampattern of the second ULA, with $H_{2,m}$, $m = 1, 2, \ldots, M_0$ being the coefficients of $\mathbf{h}_2$. Let $\mathcal{Z}_1 = e^{\jmath M_0 \varpi(\theta)}$ and $\mathcal{Z}_2 = e^{\jmath \varpi(\theta)}$, we can express the global beampattern as a polynomial in two variables, which is the product of two polynomials (of degree $M_0 - 1$) in one variable each, i.e.,

$$\begin{aligned}\mathcal{B}\left(\mathcal{Z}_1, \mathcal{Z}_2\right) &= \mathcal{B}_1\left(\mathcal{Z}_1\right) \times \mathcal{B}_2\left(\mathcal{Z}_2\right) \\ &= \left(\sum_{m_1=1}^{M_0} H_{1,m_1} \mathcal{Z}_1^{m_1-1} \right) \left(\sum_{m_2=1}^{M_0} H_{2,m_2} \mathcal{Z}_2^{m_2-1} \right). \end{aligned} \tag{2.23}$$

From this perspective, we can see that this beampattern has at most $2(M_0 - 1)$ distinct nulls (between 0 and π), while the beampattern with the conventional approach has at most $M_0^2 - 1$ distinct nulls (between 0 and π). The fact that the global beampattern, $\mathcal{B}_\theta\left(\mathbf{h}\right)$, can be expressed as the product of two beamformer beampatterns is an interesting property that can be exploited in the design of very flexible global beamformers.

Given that the first sensor is the reference, we can define the input signal-to-noise ratio (SNR) with respect to this reference as

$$\mathrm{iSNR} = \frac{\phi_X}{\phi_{V_1}}. \tag{2.24}$$

The output SNR is defined [from the variance of Z, see (2.18)] as

$$\begin{aligned}\mathrm{oSNR}\left(\mathbf{h}\right) &= \phi_X \frac{\left|\mathbf{h}^H \mathbf{d}_{\theta_\mathrm{d}}\right|^2}{\mathbf{h}^H \mathbf{\Phi}_\mathbf{v} \mathbf{h}} \\ &= \frac{\phi_X}{\phi_{V_1}} \times \frac{\left|\mathbf{h}^H \mathbf{d}_{\theta_\mathrm{d}}\right|^2}{\mathbf{h}^H \mathbf{\Gamma}_\mathbf{v} \mathbf{h}}. \end{aligned} \tag{2.25}$$

The definition of the gain in SNR is obtained from the previous definitions, i.e.,

$$\begin{aligned}\mathcal{G}\left(\mathbf{h}\right) &= \frac{\mathrm{oSNR}\left(\mathbf{h}\right)}{\mathrm{iSNR}} \\ &= \frac{\left|\mathbf{h}^H \mathbf{d}_{\theta_\mathrm{d}}\right|^2}{\mathbf{h}^H \mathbf{\Gamma}_\mathbf{v} \mathbf{h}}. \end{aligned} \tag{2.26}$$

One convenient way to evaluate the sensitivity of the global ULA to some of its imperfections is via the so-called white noise gain (WNG), which is defined by taking $\mathbf{\Gamma}_\mathbf{v} = \mathbf{I}_M$ in (2.26), where $\mathbf{I}_M$ is the $M \times M$ identity matrix, i.e.,

$$\begin{aligned}\mathcal{W}(\mathbf{h}) &= \frac{\left|\mathbf{h}^H \mathbf{d}_{\theta_\mathrm{d}}\right|^2}{\mathbf{h}^H \mathbf{h}} \\ &= \frac{\left|\mathbf{h}_1^H \mathbf{d}_{1,\theta_\mathrm{d}}\right|^2}{\mathbf{h}_1^H \mathbf{h}_1} \times \frac{\left|\mathbf{h}_2^H \mathbf{d}_{2,\theta_\mathrm{d}}\right|^2}{\mathbf{h}_2^H \mathbf{h}_2} \\ &= \mathcal{W}_1(\mathbf{h}_1) \times \mathcal{W}_2(\mathbf{h}_2),\end{aligned} \tag{2.27}$$

where

$$\mathcal{W}_1(\mathbf{h}_1) = \frac{\left|\mathbf{h}_1^H \mathbf{d}_{1,\theta_\mathrm{d}}\right|^2}{\mathbf{h}_1^H \mathbf{h}_1} \tag{2.28}$$

and

$$\mathcal{W}_2(\mathbf{h}_2) = \frac{\left|\mathbf{h}_2^H \mathbf{d}_{2,\theta_\mathrm{d}}\right|^2}{\mathbf{h}_2^H \mathbf{h}_2}. \tag{2.29}$$

Obviously, the WNG of the global ULA is simply the product of the WNGs of the first and second ULAs described in Section 2.1. It is easy to check that

$$\mathcal{W}(\mathbf{h}) \leq M, \ \forall \mathbf{h}. \tag{2.30}$$

Another important measure, which quantifies how the microphone array performs in the presence of spatial acoustic noise and reverberation is the directivity factor (DF). Considering the spherically isotropic (diffuse) noise field, the DF is defined as [6]

$$\begin{aligned}\mathcal{D}(\mathbf{h}) &= \frac{\left|\mathcal{B}_{\theta_\mathrm{d}}(\mathbf{h})\right|^2}{\frac{1}{2}\int_0^\pi \left|\mathcal{B}_\theta(\mathbf{h})\right|^2 \sin\theta d\theta} \\ &= \frac{\left|\mathcal{B}_{1,\theta_\mathrm{d}}(\mathbf{h}_1)\right|^2 \left|\mathcal{B}_{2,\theta_\mathrm{d}}(\mathbf{h}_2)\right|^2}{\frac{1}{2}\int_0^\pi \left|\mathcal{B}_{1,\theta}(\mathbf{h}_1)\right|^2 \left|\mathcal{B}_{2,\theta}(\mathbf{h}_2)\right|^2 \sin\theta d\theta} \\ &= \frac{\left|\mathbf{h}^H \mathbf{d}_{\theta_\mathrm{d}}\right|^2}{\mathbf{h}^H \mathbf{\Gamma} \mathbf{h}},\end{aligned} \tag{2.31}$$

where $\mathbf{\Gamma}$ is defined in (2.11). It is clear that

$$\mathcal{D}(\mathbf{h}) \leq \mathbf{d}_{\theta_\mathrm{d}}^H \mathbf{\Gamma}^{-1} \mathbf{d}_{\theta_\mathrm{d}}, \ \forall \mathbf{h}. \tag{2.32}$$

We observe that contrary to the beampattern and the WNG, the DF of the global ULA cannot be factorized, i.e.,

$$\mathcal{D}(\mathbf{h}) \neq \mathcal{D}_1(\mathbf{h}_1) \times \mathcal{D}_2(\mathbf{h}_2), \tag{2.33}$$

where

$$\mathcal{D}_1(\mathbf{h}_1) = \frac{|\mathcal{B}_{1,\theta_\mathrm{d}}(\mathbf{h}_1)|^2}{\frac{1}{2}\int_0^\pi |\mathcal{B}_{1,\theta}(\mathbf{h}_1)|^2 \sin\theta d\theta} = \frac{\left|\mathbf{h}_1^H \mathbf{d}_{1,\theta_\mathrm{d}}\right|^2}{\mathbf{h}_1^H \mathbf{\Gamma}_1 \mathbf{h}_1}, \tag{2.34}$$

$$\mathcal{D}_2(\mathbf{h}_2) = \frac{|\mathcal{B}_{2,\theta_\mathrm{d}}(\mathbf{h}_2)|^2}{\frac{1}{2}\int_0^\pi |\mathcal{B}_{2,\theta}(\mathbf{h}_2)|^2 \sin\theta d\theta} = \frac{\left|\mathbf{h}_2^H \mathbf{d}_{2,\theta_\mathrm{d}}\right|^2}{\mathbf{h}_2^H \mathbf{\Gamma}_2 \mathbf{h}_2}, \tag{2.35}$$

with

$$\mathbf{\Gamma}_1 = \frac{1}{2}\int_0^\pi \mathbf{d}_{1,\theta}\mathbf{d}_{1,\theta}^H \sin\theta d\theta, \tag{2.36}$$

$$\mathbf{\Gamma}_2 = \frac{1}{2}\int_0^\pi \mathbf{d}_{2,\theta}\mathbf{d}_{2,\theta}^H \sin\theta d\theta. \tag{2.37}$$

The elements of the $M_0 \times M_0$ matrices $\mathbf{\Gamma}_1(\omega)$ and $\mathbf{\Gamma}_2(\omega)$ are given, respectively, by

$$[\mathbf{\Gamma}_1(\omega)]_{ij} = \mathrm{sinc}\left[\omega(j-i)M_0\delta/c\right] \tag{2.38}$$

and

$$[\mathbf{\Gamma}_2(\omega)]_{ij} = \mathrm{sinc}\left[\omega(j-i)\delta/c\right], \tag{2.39}$$

with $[\mathbf{\Gamma}_1(\omega)]_{mm} = [\mathbf{\Gamma}_2(\omega)]_{mm} = 1,\ m = 1, 2, \ldots, M_0$. In fact, one can verify that $\mathbf{\Gamma} \neq \mathbf{\Gamma}_1 \otimes \mathbf{\Gamma}_2$, that is why (2.33) is true in general.

One can check that

$$\mathbf{h}_1 \otimes \mathbf{h}_2 = (\mathbf{h}_1 \otimes \mathbf{I}_{M_0})\,\mathbf{h}_2 \tag{2.40}$$

$$= (\mathbf{I}_{M_0} \otimes \mathbf{h}_2)\,\mathbf{h}_1, \tag{2.41}$$

where $\mathbf{I}_{M_0}$ is the $M_0 \times M_0$ identity matrix. Basically, the previous expressions, which separate $\mathbf{h}_2$ and $\mathbf{h}_1$ into matrix-vector products, will often be very helpful and convenient to use in the derivation of beamformers.

When $\mathbf{h}_2$ is fixed, given, and satisfies the distortionless constraint, i.e., $\mathbf{h}_2^H \mathbf{d}_{2,\theta_\mathrm{d}} = 1$; then, thanks to (2.41), we can write the DF as

$$\mathcal{D}(\mathbf{h}_1|\mathbf{h}_2) = \frac{|\mathcal{B}_{1,\theta_d}(\mathbf{h}_1)|^2}{\mathbf{h}_1^H \mathbf{\Gamma}_{\mathbf{h}_2} \mathbf{h}_1} = \frac{\left|\mathbf{h}_1^H \mathbf{d}_{1,\theta_d}\right|^2}{\mathbf{h}_1^H \mathbf{\Gamma}_{\mathbf{h}_2} \mathbf{h}_1}, \tag{2.42}$$

where

$$\mathbf{\Gamma}_{\mathbf{h}_2} = \frac{1}{2}\int_0^{\pi} \mathbf{d}_{1,\theta}\mathbf{d}_{1,\theta}^H |\mathcal{B}_{2,\theta}(\mathbf{h}_2)|^2 \sin\theta d\theta = (\mathbf{I}_{M_0} \otimes \mathbf{h}_2)^H \mathbf{\Gamma} (\mathbf{I}_{M_0} \otimes \mathbf{h}_2). \tag{2.43}$$

In the same way, when $\mathbf{h}_1$ is fixed, given, and satisfies the distortionless constraint, i.e., $\mathbf{h}_1^H \mathbf{d}_{1,\theta_d} = 1$; then, thanks to (2.40), we can express the DF as

$$\mathcal{D}(\mathbf{h}_2|\mathbf{h}_1) = \frac{|\mathcal{B}_{2,\theta_d}(\mathbf{h}_2)|^2}{\mathbf{h}_2^H \mathbf{\Gamma}_{\mathbf{h}_1} \mathbf{h}_2} = \frac{\left|\mathbf{h}_2^H \mathbf{d}_{2,\theta_d}\right|^2}{\mathbf{h}_2^H \mathbf{\Gamma}_{\mathbf{h}_1} \mathbf{h}_2}, \tag{2.44}$$

where

$$\mathbf{\Gamma}_{\mathbf{h}_1} = \frac{1}{2}\int_0^{\pi} \mathbf{d}_{2,\theta}\mathbf{d}_{2,\theta}^H |\mathcal{B}_{1,\theta}(\mathbf{h}_1)|^2 \sin\theta d\theta = (\mathbf{h}_1 \otimes \mathbf{I}_{M_0})^H \mathbf{\Gamma} (\mathbf{h}_1 \otimes \mathbf{I}_{M_0}). \tag{2.45}$$

References

1. B. D. Van Veen and K. M. Buckley, "Beamforming: a versatile approach to spatial filtering," *IEEE Acoust., Speech, Signal Process. Mag.*, vol. 5, pp. 4–24, Apr. 1988.
2. D. H. Johnson and D. E. Dudgeon, *Array Signal Processing: Concepts and Techniques.* Signal Processing Series. Englewood Cliffs, NJ: Prentice-Hall, 1993.
3. J. Benesty, J. Chen, and Y. Huang, *Microphone Array Signal Processing.* Berlin, Germany: Springer-Verlag, 2008.
4. J. Benesty, I. Cohen, and J. Chen, *Fundamentals of Signal Enhancement and Array Signal Processing.* Singapore: Wiley–IEEE Press, 2018.
5. Y. I. Abramovich, G. J. Frazer, and B. A. Johnson, "Iterative adaptive Kronecker MIMO radar beamformer: description and convergence analysis," *IEEE Trans. Signal Process.*, vol. 58, pp. 3681–3691, Jul. 2010.
6. H. L. Van Trees, *Optimum Array Processing: Part IV of Detection, Estimation, and Modulation Theory.* New York, NY: John Wiley & Sons, Inc., 2002.

Chapter 3
Beamforming with Uniform Linear Arrays

Any good microphone array system requires a reliable beamforming algorithm at the outputs of the sensors to enhance a desired signal coming from a known direction. There are many ways to optimize the coefficients of this beamformer depending on what we want and the application at hand. Fundamentally, there are three large classes of conventional beamformers; they are the fixed, adaptive, and differential beamformers. In this chapter, we show how to derive most of their counterparts as well as new approaches with Kronecker product filters. We also show how to combine fixed and adaptive beamforming. While most of these beamformers are rather easy to derive, for some it is necessary to use iterative algorithms. The focus is on ULAs and with the decomposition of the steering vector of Chapter 2.

3.1 Fixed Beamformers

In this first section, we derive many examples of fixed beamformers thanks to the Kronecker product decomposition. We start with the most obvious one.

3.1.1 Delay and Sum

The most well-known and popular fixed beamformer is the so-called delay and sum (DS), which is derived by maximizing the WNG. Given the structure of the WNG of $\mathbf{h}$, it is clear that the maximization of this gain is equivalent to maximizing $\mathcal{W}_1(\mathbf{h}_1)$ and $\mathcal{W}_2(\mathbf{h}_2)$ separately. Taking into account the distortionless constraints, we easily get the DS beamformers at the two subarrays:

J. Benesty et al., *Array Processing*, Springer Topics in Signal Processing 18, https://doi.org/10.1007/978-3-030-15600-8_3

$$\mathbf{h}_{1,\mathrm{DS}} = \frac{\mathbf{d}_{1,\theta_\mathrm{d}}}{M_0}, \tag{3.1}$$

$$\mathbf{h}_{2,\mathrm{DS}} = \frac{\mathbf{d}_{2,\theta_\mathrm{d}}}{M_0}. \tag{3.2}$$

As a consequence, the DS beamformer corresponding to the global ULA is

$$\begin{aligned}\mathbf{h}_{\mathrm{DS}} &= \mathbf{h}_{1,\mathrm{DS}} \otimes \mathbf{h}_{2,\mathrm{DS}} \\ &= \frac{\mathbf{d}_{1,\theta_\mathrm{d}} \otimes \mathbf{d}_{2,\theta_\mathrm{d}}}{M_0^2} \\ &= \frac{\mathbf{d}_{\theta_\mathrm{d}}}{M_0^2},\end{aligned} \tag{3.3}$$

which is, of course, the classical DS beamformer [1], [2]. Here, however, it is shown how the structure of the global steering vector is exploited. In other words, the DS beamformer is determined by $2M_0$ different coefficients only when $M = M_0^2$.

It is obvious that

$$\mathcal{W}(\mathbf{h}_{\mathrm{DS}}) = M_0^2 = M \tag{3.4}$$

and the beampattern of the DS beamformer is

$$\begin{aligned}\mathcal{B}_\theta(\mathbf{h}_{\mathrm{DS}}) &= \mathcal{B}_{1,\theta}(\mathbf{h}_{1,\mathrm{DS}}) \times \mathcal{B}_{2,\theta}(\mathbf{h}_{2,\mathrm{DS}}) \\ &= \frac{1}{M_0^2}\left(\mathbf{d}_{1,\theta}^H \mathbf{d}_{1,\theta_\mathrm{d}}\right)\left(\mathbf{d}_{2,\theta}^H \mathbf{d}_{2,\theta_\mathrm{d}}\right).\end{aligned} \tag{3.5}$$

Finally, the DF of $\mathbf{h}_{\mathrm{DS}}$ is

$$\mathcal{D}(\mathbf{h}_{\mathrm{DS}}) = \frac{M_0^4}{\mathbf{d}_{\theta_\mathrm{d}}^H \mathbf{\Gamma} \mathbf{d}_{\theta_\mathrm{d}}}. \tag{3.6}$$

3.1.2 Partial Maximum DF

There are different fixed beamformers for which the DF is only maximized in part. We review some possibilities.

In the first approach, we assume that $\mathbf{h}_2$ is fixed. We may take $\mathbf{h}_2 = \mathbf{h}_{2,\mathrm{DS}}$ for the second ULA. Substituting this filter into (2.42), we get

$$\mathcal{D}(\mathbf{h}_1|\mathbf{h}_{2,\mathrm{DS}}) = \frac{\left|\mathbf{h}_1^H \mathbf{d}_{1,\theta_\mathrm{d}}\right|^2}{\mathbf{h}_1^H \mathbf{\Gamma}_{\mathbf{h}_{2,\mathrm{DS}}} \mathbf{h}_1}, \tag{3.7}$$

where

$$\mathbf{\Gamma}_{\mathbf{h}_{2,\mathrm{DS}}} = \left(\mathbf{I}_{M_0} \otimes \mathbf{h}_{2,\mathrm{DS}}\right)^H \mathbf{\Gamma} \left(\mathbf{I}_{M_0} \otimes \mathbf{h}_{2,\mathrm{DS}}\right). \tag{3.8}$$

The maximization of $\mathcal{D}(\mathbf{h}_1|\mathbf{h}_{2,\mathrm{DS}})$ gives the maximum DF beamformer at the first ULA:

$$\mathbf{h}_{1,\mathrm{mDF1}} = \frac{\mathbf{\Gamma}_{\mathbf{h}_{2,\mathrm{DS}}}^{-1} \mathbf{d}_{1,\theta_\mathrm{d}}}{\mathbf{d}_{1,\theta_\mathrm{d}}^H \mathbf{\Gamma}_{\mathbf{h}_{2,\mathrm{DS}}}^{-1} \mathbf{d}_{1,\theta_\mathrm{d}}}. \tag{3.9}$$

Therefore, our first (global) partial maximum DF (PmDF) beamformer is

$$\mathbf{h}_{\mathrm{PmDF1}} = \mathbf{h}_{1,\mathrm{mDF1}} \otimes \mathbf{h}_{2,\mathrm{DS}}. \tag{3.10}$$

We deduce that the WNG and the beampattern are, respectively,

$$\mathcal{W}(\mathbf{h}_{\mathrm{PmDF1}}) = M_0 \mathcal{W}(\mathbf{h}_{1,\mathrm{mDF1}}) \tag{3.11}$$

and

$$\mathcal{B}_\theta(\mathbf{h}_{\mathrm{PmDF1}}) = \mathcal{B}_{1,\theta}(\mathbf{h}_{1,\mathrm{mDF1}}) \times \mathcal{B}_{2,\theta}(\mathbf{h}_{2,\mathrm{DS}}). \tag{3.12}$$

Figure 3.1 displays the directivity patterns of the first partial maximum DF beamformer, $\mathbf{h}_{\mathrm{PmDF1}}$, for $\theta_\mathrm{d} = 0$, $f = 1$ kHz, $\delta = 1$ cm, and different numbers of sensors $M = M_0^2$. Figure 3.2 shows plots of the DFs and WNGs of the first partial maximum DF beamformer as a function of frequency for $\theta_\mathrm{d} = 0$, $\delta = 1$ cm, and different numbers of sensors. We observe that as the number of sensors increases, the width of the main beam and the level of side lobes decrease, while the DF of the first partial maximum DF beamformer increases. However, using a larger number of sensors increases the WNG of the first partial maximum DF beamformer only for high frequencies, but decreases its WNG for low frequencies.

In the second approach, we assume that $\mathbf{h}_1$ is fixed, i.e., $\mathbf{h}_1 = \mathbf{h}_{1,\mathrm{DS}}$ for the first ULA. Substituting this filter into (2.44), we get

$$\mathcal{D}(\mathbf{h}_2|\mathbf{h}_{1,\mathrm{DS}}) = \frac{\left|\mathbf{h}_2^H \mathbf{d}_{2,\theta_\mathrm{d}}\right|^2}{\mathbf{h}_2^H \mathbf{\Gamma}_{\mathbf{h}_{1,\mathrm{DS}}} \mathbf{h}_2}, \tag{3.13}$$

where

$$\mathbf{\Gamma}_{\mathbf{h}_{1,\mathrm{DS}}} = \left(\mathbf{h}_{1,\mathrm{DS}} \otimes \mathbf{I}_{M_0}\right)^H \mathbf{\Gamma} \left(\mathbf{h}_{1,\mathrm{DS}} \otimes \mathbf{I}_{M_0}\right). \tag{3.14}$$

The maximization of $\mathcal{D}(\mathbf{h}_2|\mathbf{h}_{1,\mathrm{DS}})$ gives the maximum DF beamformer at the second ULA:

$$\mathbf{h}_{2,\mathrm{mDF2}} = \frac{\mathbf{\Gamma}_{\mathbf{h}_{1,\mathrm{DS}}}^{-1} \mathbf{d}_{2,\theta_\mathrm{d}}}{\mathbf{d}_{2,\theta_\mathrm{d}}^H \mathbf{\Gamma}_{\mathbf{h}_{1,\mathrm{DS}}}^{-1} \mathbf{d}_{2,\theta_\mathrm{d}}}. \tag{3.15}$$

As a result, our second partial maximum DF beamformer is

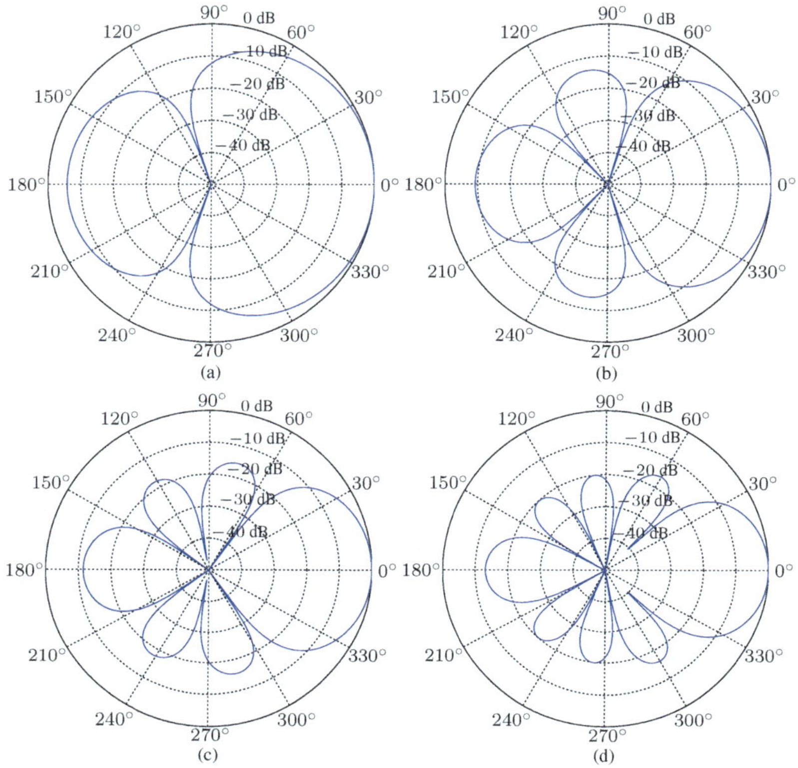

Fig. 3.1 Beampatterns of the first partial maximum DF beamformer, $\mathbf{h}_{\mathrm{PmDF1}}$, for $\theta_{\mathrm{d}} = 0$, $f = 1$ kHz, $\delta = 1$ cm, and different numbers of sensors $M = M_0^2$: (a) $M_0 = 2$, (b) $M_0 = 3$, (c) $M_0 = 4$, and (d) $M_0 = 5$.

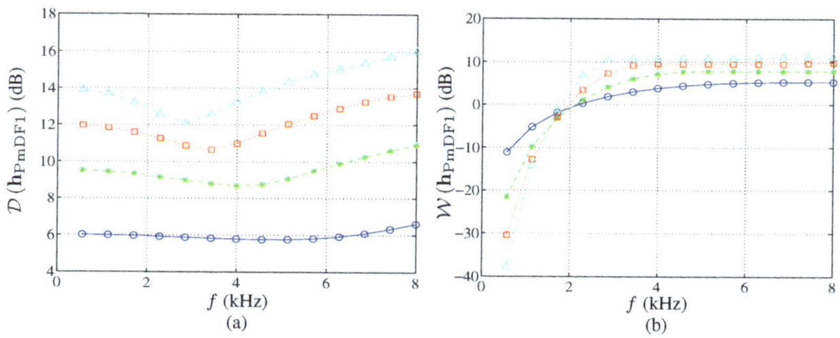

Fig. 3.2 Performance of the first partial maximum DF beamformer, $\mathbf{h}_{\mathrm{PmDF1}}$, as a function of frequency for $\theta_{\mathrm{d}} = 0$, $\delta = 1$ cm, and different numbers of sensors $M = M_0^2$: $M_0 = 2$ (solid line with circles), $M_0 = 3$ (dashed line with asterisks), $M_0 = 4$ (dotted line with squares), and $M_0 = 5$ (dash-dot line with triangles). (a) DF and (b) WNG.

$$\mathbf{h}_{\mathrm{PmDF2}} = \mathbf{h}_{1,\mathrm{DS}} \otimes \mathbf{h}_{2,\mathrm{mDF2}}. \tag{3.16}$$

We deduce that the WNG and the beampattern are, respectively,

$$\mathcal{W}\left(\mathbf{h}_{\mathrm{PmDF2}}\right) = M_0 \mathcal{W}\left(\mathbf{h}_{2,\mathrm{mDF2}}\right) \tag{3.17}$$

and

$$\mathcal{B}_{\theta}\left(\mathbf{h}_{\mathrm{PmDF2}}\right) = \mathcal{B}_{1,\theta}\left(\mathbf{h}_{1,\mathrm{DS}}\right) \times \mathcal{B}_{2,\theta}\left(\mathbf{h}_{2,\mathrm{mDF2}}\right). \tag{3.18}$$

From the two maximum DF beamformers derived above for the two subarrays, we find the third approach:

$$\mathbf{h}_{\mathrm{PmDF3}} = \mathbf{h}_{1,\mathrm{mDF1}} \otimes \mathbf{h}_{2,\mathrm{mDF2}}. \tag{3.19}$$

Now, we can maximize separately the two DFs, $\mathcal{D}_1\left(\mathbf{h}_1\right)$ and $\mathcal{D}_2\left(\mathbf{h}_2\right)$, of the subarrays. We get

$$\mathbf{h}_{1,\mathrm{mDF}} = \frac{\mathbf{\Gamma}_1^{-1}\mathbf{d}_{1,\theta_{\mathrm{d}}}}{\mathbf{d}_{1,\theta_{\mathrm{d}}}^H \mathbf{\Gamma}_1^{-1}\mathbf{d}_{1,\theta_{\mathrm{d}}}}, \tag{3.20}$$

$$\mathbf{h}_{2,\mathrm{mDF}} = \frac{\mathbf{\Gamma}_2^{-1}\mathbf{d}_{2,\theta_{\mathrm{d}}}}{\mathbf{d}_{2,\theta_{\mathrm{d}}}^H \mathbf{\Gamma}_2^{-1}\mathbf{d}_{2,\theta_{\mathrm{d}}}}. \tag{3.21}$$

As a result, the partial maximum DF beamformer of the fourth approach is simply the Kronecker product of the two above filters, i.e.,

$$\mathbf{h}_{\mathrm{PmDF4}} = \mathbf{h}_{1,\mathrm{mDF}} \otimes \mathbf{h}_{2,\mathrm{mDF}}. \tag{3.22}$$

Figure 3.3 displays the directivity patterns of the fourth partial maximum DF beamformer, $\mathbf{h}_{\mathrm{PmDF4}}$, for $\theta_{\mathrm{d}} = 0$, $f = 1$ kHz, $\delta = 1$ cm, and different numbers of sensors $M = M_0^2$. Figure 3.4 shows plots of the DFs and WNGs of the fourth partial maximum DF beamformer as a function of frequency for $\theta_{\mathrm{d}} = 0$, $\delta = 1$ cm, and different numbers of sensors. We observe that as the number of sensors increases, the width of the main beam and the level of side lobes decrease, while the DF of the fourth partial maximum DF beamformer increases. However, using a larger number of sensors decreases the WNG of the fourth partial maximum DF beamformer, especially at low frequencies. For a given number of sensors, the DF of the fourth partial maximum DF beamformer is higher than that of the first partial maximum DF beamformer, but the WNG of the fourth partial maximum DF beamformer is lower than that of the first partial maximum DF beamformer (compare Figs 3.2 and 3.4).

Two other possibilities are

$$\mathbf{h}_{\mathrm{PmDF5}} = \mathbf{h}_{1,\mathrm{mDF}} \otimes \mathbf{h}_{2,\mathrm{DS}} \tag{3.23}$$

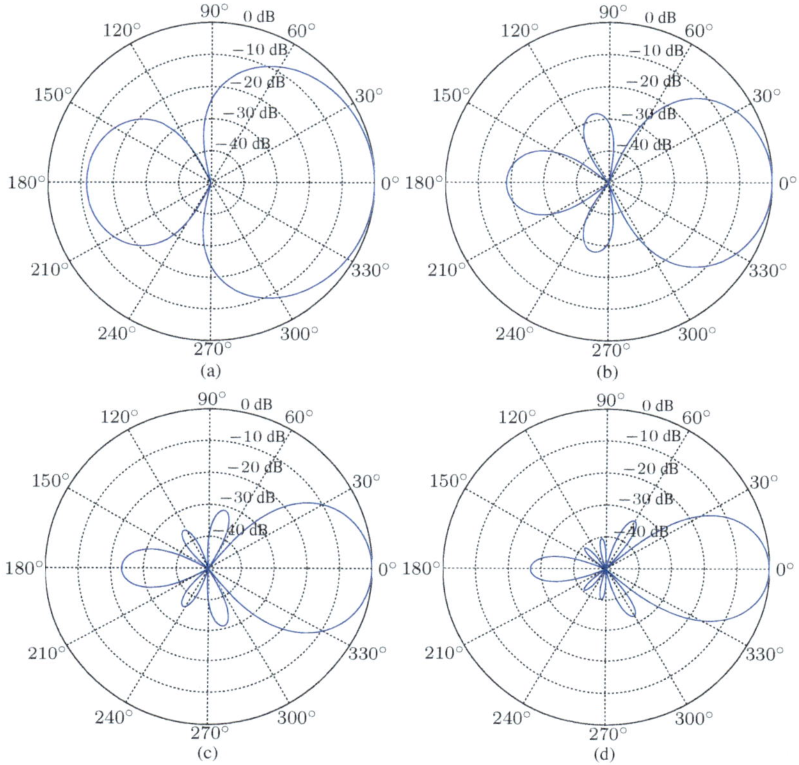

Fig. 3.3 Beampatterns of the fourth partial maximum DF beamformer, $\mathbf{h}_{\text{PmDF4}}$, for $\theta_{\text{d}} = 0$, $f = 1$ kHz, $\delta = 1$ cm, and different numbers of sensors $M = M_0^2$: (a) $M_0 = 2$, (b) $M_0 = 3$, (c) $M_0 = 4$, and (d) $M_0 = 5$.

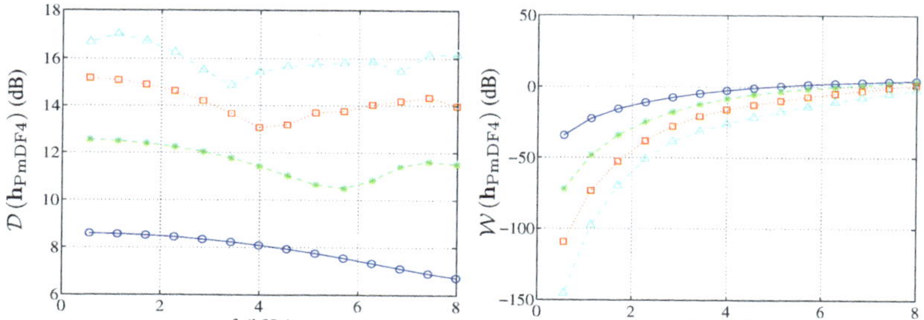

Fig. 3.4 Performance of the fourth partial maximum DF beamformer, $\mathbf{h}_{\text{PmDF4}}$, as a function of frequency for $\theta_{\text{d}} = 0$, $\delta = 1$ cm, and different numbers of sensors $M = M_0^2$: $M_0 = 2$ (solid line with circles), $M_0 = 3$ (dashed line with asterisks), $M_0 = 4$ (dotted line with squares), and $M_0 = 5$ (dash-dot line with triangles). (a) DF and (b) WNG.

and

$$\mathbf{h}_{\mathrm{PmDF6}} = \mathbf{h}_{1,\mathrm{DS}} \otimes \mathbf{h}_{2,\mathrm{mDF}}. \tag{3.24}$$

3.1.3 Maximum DF

From a theoretical point of view, the maximum DF beamformer is obtained by maximizing $\mathcal{D}(\mathbf{h})$ in (2.31) but this DF cannot be maximized directly. Therefore, an iterative algorithm is required for this task.

We start by taking

$$\begin{aligned}\mathbf{h}_2^{(0)} &= \mathbf{h}_{2,\mathrm{mDF}} \\ &= \frac{\mathbf{\Gamma}_2^{-1}\mathbf{d}_{2,\theta_\mathrm{d}}}{\mathbf{d}_{2,\theta_\mathrm{d}}^H \mathbf{\Gamma}_2^{-1}\mathbf{d}_{2,\theta_\mathrm{d}}}.\end{aligned} \tag{3.25}$$

Substituting $\mathbf{h}_2^{(0)}$ into (2.43), we get

$$\mathbf{\Gamma}_{\mathbf{h}_2^{(0)}} = \left(\mathbf{I}_{M_0} \otimes \mathbf{h}_2^{(0)}\right)^H \mathbf{\Gamma} \left(\mathbf{I}_{M_0} \otimes \mathbf{h}_2^{(0)}\right). \tag{3.26}$$

Now, plugging this expression into the DF in (2.42), we obtain at iteration 1:

$$\mathcal{D}\left(\mathbf{h}_1^{(1)} | \mathbf{h}_2^{(0)}\right) = \frac{\left|\left(\mathbf{h}_1^{(1)}\right)^H \mathbf{d}_{1,\theta_\mathrm{d}}\right|^2}{\left(\mathbf{h}_1^{(1)}\right)^H \mathbf{\Gamma}_{\mathbf{h}_2^{(0)}} \mathbf{h}_1^{(1)}}. \tag{3.27}$$

The maximization of $\mathcal{D}\left(\mathbf{h}_1^{(1)} | \mathbf{h}_2^{(0)}\right)$ with respect to $\mathbf{h}_1^{(1)}$ gives

$$\mathbf{h}_1^{(1)} = \frac{\mathbf{\Gamma}_{\mathbf{h}_2^{(0)}}^{-1} \mathbf{d}_{1,\theta_\mathrm{d}}}{\mathbf{d}_{1,\theta_\mathrm{d}}^H \mathbf{\Gamma}_{\mathbf{h}_2^{(0)}}^{-1} \mathbf{d}_{1,\theta_\mathrm{d}}}. \tag{3.28}$$

Using $\mathbf{h}_1^{(1)}$ in (2.45), we get

$$\mathbf{\Gamma}_{\mathbf{h}_1^{(1)}} = \left(\mathbf{h}_1^{(1)} \otimes \mathbf{I}_{M_0}\right)^H \mathbf{\Gamma} \left(\mathbf{h}_1^{(1)} \otimes \mathbf{I}_{M_0}\right). \tag{3.29}$$

As a result, the DF in (2.44) is

$$\mathcal{D}\left(\mathbf{h}_2^{(1)}|\mathbf{h}_1^{(1)}\right) = \frac{\left|\left(\mathbf{h}_2^{(1)}\right)^H \mathbf{d}_{2,\theta_\mathrm{d}}\right|^2}{\left(\mathbf{h}_2^{(1)}\right)^H \boldsymbol{\Gamma}_{\mathbf{h}_1^{(1)}} \mathbf{h}_2^{(1)}}, \tag{3.30}$$

whose maximization with respect to $\mathbf{h}_2^{(1)}$ gives

$$\mathbf{h}_2^{(1)} = \frac{\boldsymbol{\Gamma}_{\mathbf{h}_1^{(1)}}^{-1} \mathbf{d}_{2,\theta_\mathrm{d}}}{\mathbf{d}_{2,\theta_\mathrm{d}}^H \boldsymbol{\Gamma}_{\mathbf{h}_1^{(1)}}^{-1} \mathbf{d}_{2,\theta_\mathrm{d}}}. \tag{3.31}$$

Continuing the iterations up to the iteration n, we easily get for the first filter:

$$\mathbf{h}_1^{(n)} = \frac{\boldsymbol{\Gamma}_{\mathbf{h}_2^{(n-1)}}^{-1} \mathbf{d}_{1,\theta_\mathrm{d}}}{\mathbf{d}_{1,\theta_\mathrm{d}}^H \boldsymbol{\Gamma}_{\mathbf{h}_2^{(n-1)}}^{-1} \mathbf{d}_{1,\theta_\mathrm{d}}}, \tag{3.32}$$

with

$$\boldsymbol{\Gamma}_{\mathbf{h}_2^{(n-1)}} = \left(\mathbf{I}_{M_0} \otimes \mathbf{h}_2^{(n-1)}\right)^H \boldsymbol{\Gamma} \left(\mathbf{I}_{M_0} \otimes \mathbf{h}_2^{(n-1)}\right), \tag{3.33}$$

and for the second filter:

$$\mathbf{h}_2^{(n)} = \frac{\boldsymbol{\Gamma}_{\mathbf{h}_1^{(n)}}^{-1} \mathbf{d}_{2,\theta_\mathrm{d}}}{\mathbf{d}_{2,\theta_\mathrm{d}}^H \boldsymbol{\Gamma}_{\mathbf{h}_1^{(n)}}^{-1} \mathbf{d}_{2,\theta_\mathrm{d}}}, \tag{3.34}$$

with

$$\boldsymbol{\Gamma}_{\mathbf{h}_1^{(n)}} = \left(\mathbf{h}_1^{(n)} \otimes \mathbf{I}_{M_0}\right)^H \boldsymbol{\Gamma} \left(\mathbf{h}_1^{(n)} \otimes \mathbf{I}_{M_0}\right). \tag{3.35}$$

Finally, we deduce that the maximum DF beamformer is at iteration n:

$$\mathbf{h}_{\mathrm{mDF}}^{(n)} = \mathbf{h}_1^{(n)} \otimes \mathbf{h}_2^{(n)}. \tag{3.36}$$

Figure 3.5 displays the directivity patterns of the maximum DF beamformer, $\mathbf{h}_{\mathrm{mDF}}^{(n)}$, for $f = 1$ kHz, $\delta = 1$ cm, and $M_0 = 3$, obtained at the iteration n for several values of n. Figure 3.6 shows plots of the DFs and WNGs of the maximum DF beamformer as a function of frequency for $\delta = 1$ cm, $M_0 = 3$, and several values of n. We observe that the DF of the maximum DF beamformer increases at each iteration, and roughly converges after five iterations, while the WNG decreases at each iteration. Compared with the partial maximum DF beamformers, the maximum DF beamformer yields higher DF, but lower WNG (compare Figs 3.4 and 3.6).

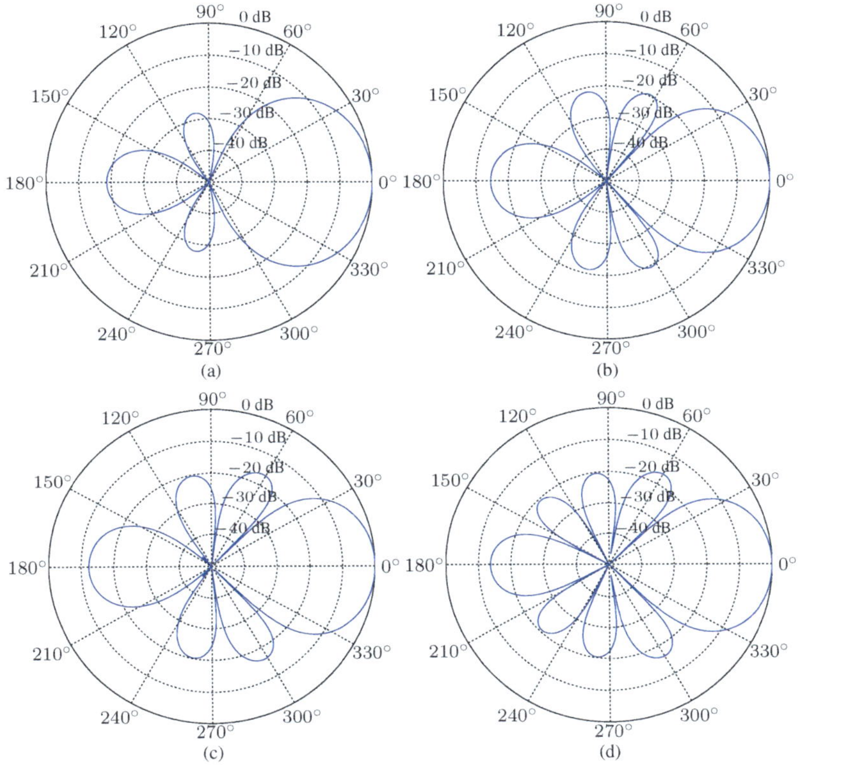

Fig. 3.5 Beampatterns of the maximum DF beamformer, $\mathbf{h}_{\mathrm{mDF}}^{(n)}$, for $f = 1$ kHz, $\delta = 1$ cm, and $M_0 = 3$, obtained at the iteration n: (a) $n = 0$, (b) $n = 1$, (c) $n = 2$, and (d) $n = 5$.

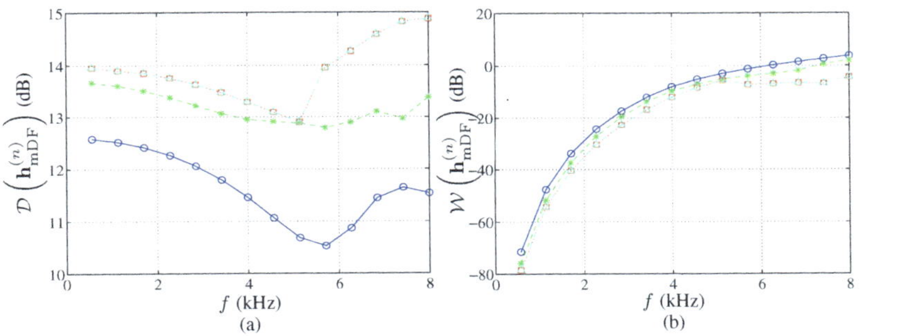

Fig. 3.6 Performance of the maximum DF beamformer, $\mathbf{h}_{\mathrm{mDF}}^{(n)}$, as a function of frequency for $\delta = 1$ cm, $M_0 = 3$, and several values of n: $n = 0$ (solid line with circles), $n = 1$ (dashed line with asterisks), $n = 5$ (dotted line with squares), and $n = 10$ (dash-dot line with triangles). (a) DF and (b) WNG.

If we want to compromise between the WNG and the DF, we should optimize the following criteria:

$$\min_{\mathbf{h}_1^{(n)}} \left(\mathbf{h}_1^{(n)}\right)^H \left(\mathbf{\Gamma}_{\mathbf{h}_2^{(n-1)}} + \epsilon_1 \mathbf{I}_{M_0}\right) \mathbf{h}_1^{(n)} \quad \text{subject to} \quad \left(\mathbf{h}_1^{(n)}\right)^H \mathbf{d}_{1,\theta_\mathrm{d}} = 1, \tag{3.37}$$

$$\min_{\mathbf{h}_2^{(n)}} \left(\mathbf{h}_2^{(n)}\right)^H \left(\mathbf{\Gamma}_{\mathbf{h}_1^{(n)}} + \epsilon_2 \mathbf{I}_{M_0}\right) \mathbf{h}_2^{(n)} \quad \text{subject to} \quad \left(\mathbf{h}_2^{(n)}\right)^H \mathbf{d}_{2,\theta_\mathrm{d}} = 1, \tag{3.38}$$

where $\epsilon_1, \epsilon_2 \geq 0$ are the regularization parameters, and $\mathbf{\Gamma}_{\mathbf{h}_2^{(n-1)}}$ and $\mathbf{\Gamma}_{\mathbf{h}_1^{(n)}}$ are defined in (3.33) and (3.35), respectively. We find that the optimal filters are

$$\mathbf{h}_{1,\epsilon_1}^{(n)} = \frac{\left(\mathbf{\Gamma}_{\mathbf{h}_2^{(n-1)}} + \epsilon_1 \mathbf{I}_{M_0}\right)^{-1} \mathbf{d}_{1,\theta_\mathrm{d}}}{\mathbf{d}_{1,\theta_\mathrm{d}}^H \left(\mathbf{\Gamma}_{\mathbf{h}_2^{(n-1)}} + \epsilon_1 \mathbf{I}_{M_0}\right)^{-1} \mathbf{d}_{1,\theta_\mathrm{d}}} \tag{3.39}$$

and

$$\mathbf{h}_{2,\epsilon_2}^{(n)} = \frac{\left(\mathbf{\Gamma}_{\mathbf{h}_1^{(n)}} + \epsilon_2 \mathbf{I}_{M_0}\right)^{-1} \mathbf{d}_{2,\theta_\mathrm{d}}}{\mathbf{d}_{2,\theta_\mathrm{d}}^H \left(\mathbf{\Gamma}_{\mathbf{h}_1^{(n)}} + \epsilon_2 \mathbf{I}_{M_0}\right)^{-1} \mathbf{d}_{2,\theta_\mathrm{d}}}, \tag{3.40}$$

with the initialization:

$$\mathbf{h}_{2,\epsilon_2}^{(0)} = \frac{\left(\mathbf{\Gamma}_2 + \epsilon_2 \mathbf{I}_{M_0}\right)^{-1} \mathbf{d}_{2,\theta_\mathrm{d}}}{\mathbf{d}_{2,\theta_\mathrm{d}}^H \left(\mathbf{\Gamma}_2 + \epsilon_2 \mathbf{I}_{M_0}\right)^{-1} \mathbf{d}_{2,\theta_\mathrm{d}}}. \tag{3.41}$$

As a result, the robust maximum DF beamformer is at iteration n:

$$\mathbf{h}_{\mathrm{R},\epsilon_1,\epsilon_2}^{(n)} = \mathbf{h}_{1,\epsilon_1}^{(n)} \otimes \mathbf{h}_{2,\epsilon_2}^{(n)}. \tag{3.42}$$

Figure 3.7 displays the directivity patterns of the robust maximum DF beamformer at the iteration $n = 5$, $\mathbf{h}_{\mathrm{R},\epsilon_1,\epsilon_2}^{(5)}$, for $M_0 = 3$, $f = 1$ kHz, $\delta = 1$ cm, and several values of ϵ_1 and ϵ_2. Figure 3.8 shows plots of the DFs and WNGs of the robust maximum DF beamformer at the iteration $n = 5$ as a function of frequency for $M_0 = 3$, $\delta = 1$ cm, and several values of ϵ_1 and ϵ_2. We observe that for larger values of ϵ_1 and ϵ_2, the WNG increases, but the DF decreases and the main lobe becomes wider.

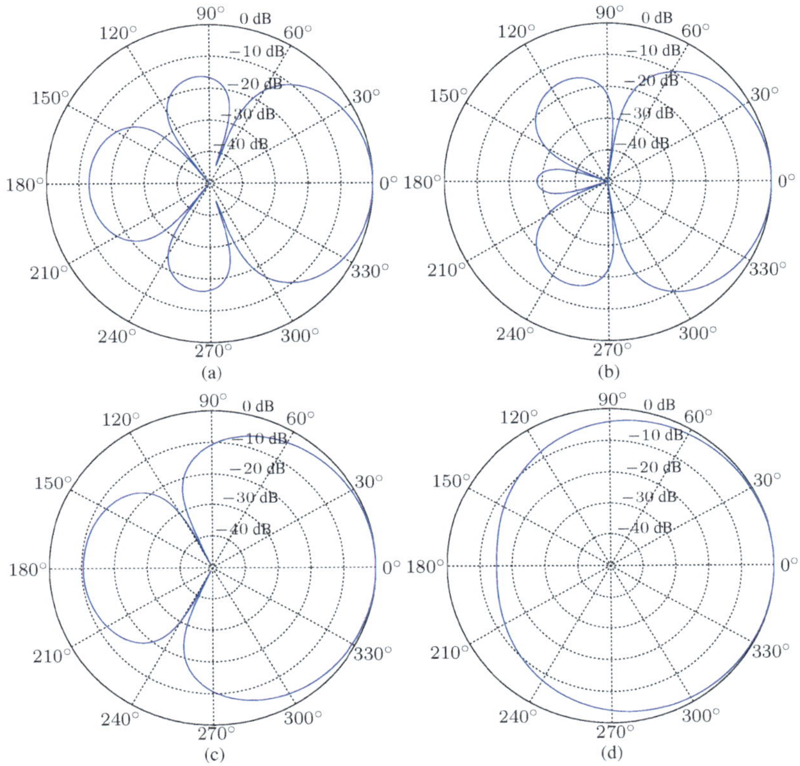

Fig. 3.7 Beampatterns of the robust maximum DF beamformer at the iteration $n = 5$, $\mathbf{h}^{(5)}_{\mathrm{R},\epsilon_1,\epsilon_2}$, for $M_0 = 3$, $f = 1$ kHz, $\delta = 1$ cm, and several values of ϵ_1 and ϵ_2: (a) $\epsilon_1 = \epsilon_2 = 0.001$, (b) $\epsilon_1 = \epsilon_2 = 0.01$, (c) $\epsilon_1 = \epsilon_2 = 0.1$, and (d) $\epsilon_1 = \epsilon_2 = 1$.

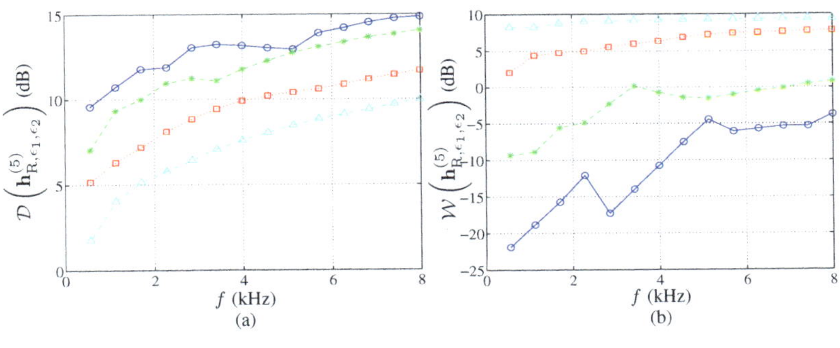

Fig. 3.8 Performance of the robust maximum DF beamformer at the iteration $n = 5$, $\mathbf{h}^{(5)}_{\mathrm{R},\epsilon_1,\epsilon_2}$, as a function of frequency for $M_0 = 3$, $\delta = 1$ cm, and several values of ϵ_1 and ϵ_2: $\epsilon_1 = \epsilon_2 = 0.001$ (solid line with circles), $\epsilon_1 = \epsilon_2 = 0.01$ (dashed line with asterisks), $\epsilon_1 = \epsilon_2 = 0.1$ (dotted line with squares), and $\epsilon_1 = \epsilon_2 = 1$ (dash-dot line with triangles). (a) DF and (b) WNG.

3.1.4 Null Steering

In this subsection, we assume that we have one interference source impinging on the global array from the direction $\theta_0 \neq \theta_{\rm d}$ that we would like to completely cancel, i.e., to steer a null in that direction, and, meanwhile, recover the desired source coming from the direction $\theta_{\rm d}$. There are many ways to do this. First, remember that a null in the beampattern $\mathcal{B}_{1,\theta}(\mathbf{h}_1)$ implies a null in the global beampattern $\mathcal{B}_\theta(\mathbf{h})$. In the same way, a null in the beampattern $\mathcal{B}_{2,\theta}(\mathbf{h}_2)$ implies a null in the global beampattern $\mathcal{B}_\theta(\mathbf{h})$. As a result, the same null in $\mathcal{B}_{1,\theta}(\mathbf{h}_1)$ and in $\mathcal{B}_{2,\theta}(\mathbf{h}_2)$ implies a null in $\mathcal{B}_\theta(\mathbf{h})$ of multiplicity 2. Then, by including the distortionless constraints, we can write the constraint equations as

$$\mathbf{C}_1^H \mathbf{h}_1 = \mathbf{i}_{\rm c}, \tag{3.43}$$

$$\mathbf{C}_2^H \mathbf{h}_2 = \mathbf{i}_{\rm c}, \tag{3.44}$$

where

$$\mathbf{C}_1 = \begin{bmatrix} \mathbf{d}_{1,\theta_{\rm d}} \ \mathbf{d}_{1,\theta_0} \end{bmatrix}, \tag{3.45}$$

$$\mathbf{C}_2 = \begin{bmatrix} \mathbf{d}_{2,\theta_{\rm d}} \ \mathbf{d}_{2,\theta_0} \end{bmatrix} \tag{3.46}$$

are the constraint matrices of size $M_0 \times 2$ whose two columns are linearly independent and

$$\mathbf{i}_{\rm c} = \begin{bmatrix} 1 \ 0 \end{bmatrix}^T \tag{3.47}$$

is a vector of length 2.

In the first approach, we take the DS beamformer for the second filter, i.e., $\mathbf{h}_2 = \mathbf{h}_{2,\rm DS}$. To find the first filter, we maximize the WNG by taking (3.43) into account, i.e.,

$$\min_{\mathbf{h}_1} \mathbf{h}_1^H \mathbf{h}_1 \quad \text{subject to} \quad \mathbf{C}_1^H \mathbf{h}_1 = \mathbf{i}_{\rm c}. \tag{3.48}$$

From this criterion, we get the minimum-norm (MN) beamformer:

$$\mathbf{h}_{1,\rm MN} = \mathbf{C}_1 \left(\mathbf{C}_1^H \mathbf{C}_1 \right)^{-1} \mathbf{i}_{\rm c}, \tag{3.49}$$

which is also the minimum-norm solution of (3.43). Therefore, the first (global) proposed null-steering (NS) beamformer is

$$\mathbf{h}_{\rm NS1} = \mathbf{h}_{1,\rm MN} \otimes \mathbf{h}_{2,\rm DS}. \tag{3.50}$$

We deduce that the WNG and the beampattern are, respectively,

$$\mathcal{W}(\mathbf{h}_{\rm NS1}) = M_0 \mathcal{W}(\mathbf{h}_{1,\rm MN}) \tag{3.51}$$

and

$$\mathcal{B}_\theta\left(\mathbf{h}_{\mathrm{NS1}}\right) = \mathcal{B}_{1,\theta}\left(\mathbf{h}_{1,\mathrm{MN}}\right) \times \mathcal{B}_{2,\theta}\left(\mathbf{h}_{2,\mathrm{DS}}\right). \tag{3.52}$$

In the second approach, we take the DS beamformer for the first filter, i.e., $\mathbf{h}_1 = \mathbf{h}_{1,\mathrm{DS}}$. To find the second filter, we maximize the WNG by taking (3.44) into account. We get

$$\mathbf{h}_{2,\mathrm{MN}} = \mathbf{C}_2\left(\mathbf{C}_2^H\mathbf{C}_2\right)^{-1}\mathbf{i}_\mathrm{c}, \tag{3.53}$$

which is also the minimum-norm solution of (3.44). Therefore, the second proposed null-steering beamformer is

$$\mathbf{h}_{\mathrm{NS2}} = \mathbf{h}_{1,\mathrm{DS}} \otimes \mathbf{h}_{2,\mathrm{MN}}. \tag{3.54}$$

We deduce that the WNG and the beampattern are, respectively,

$$\mathcal{W}\left(\mathbf{h}_{\mathrm{NS2}}\right) = M_0\mathcal{W}\left(\mathbf{h}_{2,\mathrm{MN}}\right) \tag{3.55}$$

and

$$\mathcal{B}_\theta\left(\mathbf{h}_{\mathrm{NS2}}\right) = \mathcal{B}_{1,\theta}\left(\mathbf{h}_{1,\mathrm{DS}}\right) \times \mathcal{B}_{2,\theta}\left(\mathbf{h}_{2,\mathrm{MN}}\right). \tag{3.56}$$

Finally, in the third and last approach as far as the maximization of the WNG is concerned, we propose to use the two derived minimum-norm filters, i.e.,

$$\mathbf{h}_{\mathrm{NS3}} = \mathbf{h}_{1,\mathrm{MN}} \otimes \mathbf{h}_{2,\mathrm{MN}}. \tag{3.57}$$

The global beampattern is

$$\mathcal{B}_\theta\left(\mathbf{h}_{\mathrm{NS3}}\right) = \mathcal{B}_{1,\theta}\left(\mathbf{h}_{1,\mathrm{MN}}\right) \times \mathcal{B}_{2,\theta}\left(\mathbf{h}_{2,\mathrm{MN}}\right), \tag{3.58}$$

which has a null of multiplicity 2 in the direction θ_0.

Figure 3.9 displays the directivity patterns of the third null-steering beamformer, $\mathbf{h}_{\mathrm{NS3}}$, for $\theta_\mathrm{d} = 0$, $f = 1$ kHz, $\delta = 5$ mm, $M_0 = 3$, and several values of θ_0. Figure 3.10 shows plots of the DFs and WNGs of the third null-steering beamformer as a function of frequency for $\theta_\mathrm{d} = 0$, $\delta = 5$ mm, $M_0 = 3$, and several values of θ_0. We observe that for $\theta_\mathrm{d} = 0$, the WNG of the third null-steering beamformer increases as θ_0 increases from 90° to 180°.

A second class of beamformers is obtained by maximizing the DF instead of the WNG. Let $\mathbf{h}_2 = \mathbf{h}_{2,\mathrm{DS}}$. To find the first filter, we maximize the DF by taking (3.43) into account, i.e.,

$$\min_{\mathbf{h}_1} \mathbf{h}_1^H\boldsymbol{\Gamma}_1\mathbf{h}_1 \quad \text{subject to} \quad \mathbf{C}_1^H\mathbf{h}_1 = \mathbf{i}_\mathrm{c}. \tag{3.59}$$

The solution to this problem is

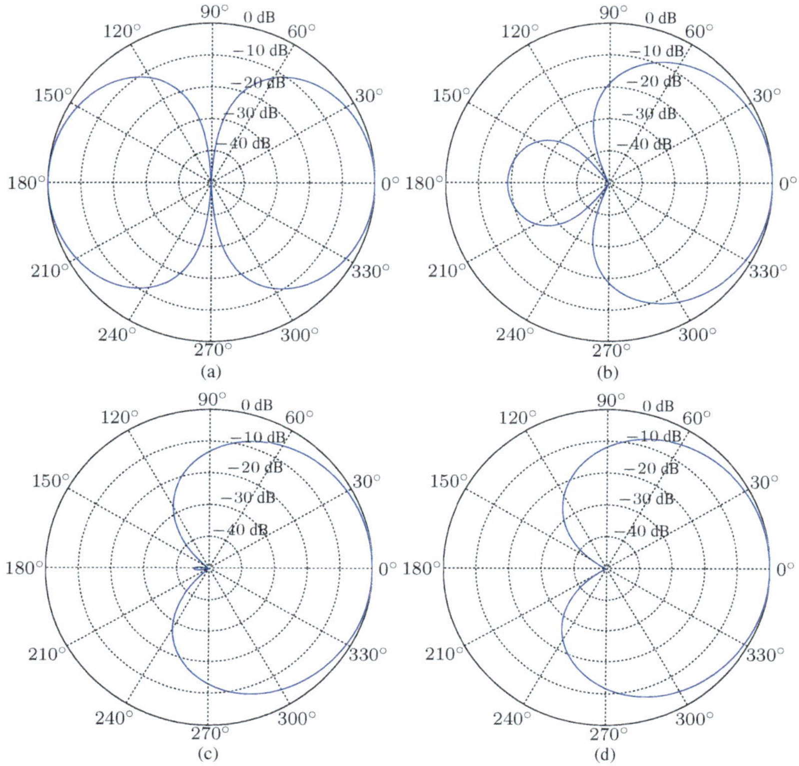

Fig. 3.9 Beampatterns of the third null-steering beamformer, $\mathbf{h}_{\mathrm{NS3}}$, for $\theta_{\mathrm{d}} = 0$, $f = 1$ kHz, $\delta = 5$ mm, $M_{\mathrm{O}} = 3$, and several values of θ_{O}: (a) $\theta_{\mathrm{O}} = 90°$, (b) $\theta_{\mathrm{O}} = 120°$, (c) $\theta_{\mathrm{O}} = 150°$, and (d) $\theta_{\mathrm{O}} = 180°$.

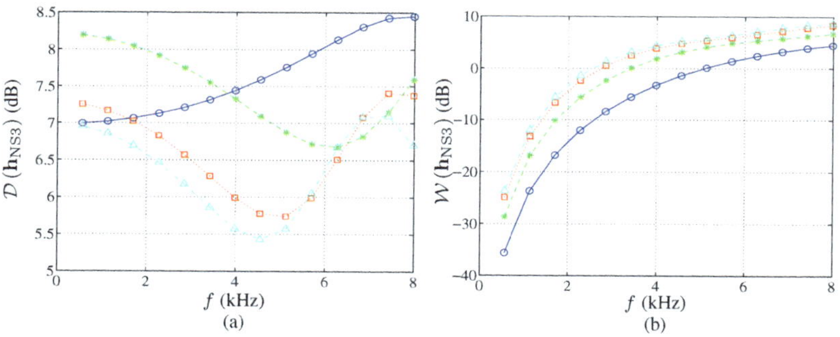

Fig. 3.10 Performance of the third null-steering beamformer, $\mathbf{h}_{\mathrm{NS3}}$, as a function of frequency for $\theta_{\mathrm{d}} = 0$, $\delta = 5$ mm, $M_{\mathrm{O}} = 3$, and several values of θ_{O}: $\theta_{\mathrm{O}} = 90°$ (solid line with circles), $\theta_{\mathrm{O}} = 120°$ (dashed line with asterisks), $\theta_{\mathrm{O}} = 150°$ (dotted line with squares), and $\theta_{\mathrm{O}} = 180°$ (dash-dot line with triangles). (a) DF and (b) WNG.

$$\mathbf{h}_{1,\mathrm{NS4}} = \mathbf{\Gamma}_1^{-1}\mathbf{C}_1\left(\mathbf{C}_1^H\mathbf{\Gamma}_1^{-1}\mathbf{C}_1\right)^{-1}\mathbf{i}_\mathrm{c}. \quad (3.60)$$

Therefore, the fourth approach is

$$\mathbf{h}_{\mathrm{NS4}} = \mathbf{h}_{1,\mathrm{NS4}} \otimes \mathbf{h}_{2,\mathrm{DS}}. \quad (3.61)$$

In the fifth approach, we choose $\mathbf{h}_1 = \mathbf{h}_{1,\mathrm{DS}}$ and to find the second filter, we maximize the DF by taking (3.44) into account. We get

$$\mathbf{h}_{2,\mathrm{NS5}} = \mathbf{\Gamma}_2^{-1}\mathbf{C}_2\left(\mathbf{C}_2^H\mathbf{\Gamma}_2^{-1}\mathbf{C}_2\right)^{-1}\mathbf{i}_\mathrm{c}. \quad (3.62)$$

Therefore, the global null-steering beamformer is

$$\mathbf{h}_{\mathrm{NS5}} = \mathbf{h}_{1,\mathrm{DS}} \otimes \mathbf{h}_{2,\mathrm{NS5}}. \quad (3.63)$$

In the sixth approach, we combine the two previous ones:

$$\mathbf{h}_{\mathrm{NS6}} = \mathbf{h}_{1,\mathrm{NS4}} \otimes \mathbf{h}_{2,\mathrm{NS5}}. \quad (3.64)$$

The DF of $\mathbf{h}_{\mathrm{NS6}}$ will be greater than that of $\mathbf{h}_{\mathrm{NS4}}$ and $\mathbf{h}_{\mathrm{NS5}}$.

Figure 3.11 displays the directivity patterns of the sixth null-steering beamformer, $\mathbf{h}_{\mathrm{NS6}}$, for $\theta_\mathrm{d} = 0$, $f = 1$ kHz, $\delta = 5$ mm, $M_0 = 3$, and several values of θ_0. Figure 3.12 shows plots of the DFs and WNGs of the sixth null-steering beamformer as a function of frequency for $\theta_\mathrm{d} = 0$, $\delta = 5$ mm, $M_0 = 3$, and several values of θ_0. Compared with the third null-steering beamformer, the sixth null-steering beamformer yields higher DF, but lower WNG (compare Figs 3.10 and 3.12).

To fully maximize the DF while having a null of multiplicity 2 in the direction θ_0, we need to optimize the following criteria:

$$\min_{\mathbf{h}_1^{(n)}} \left(\mathbf{h}_1^{(n)}\right)^H \mathbf{\Gamma}_{\mathbf{h}_2^{(n-1)}}\mathbf{h}_1^{(n)} \quad \text{subject to} \quad \mathbf{C}_1^H\mathbf{h}_1^{(n)} = \mathbf{i}_\mathrm{c}, \quad (3.65)$$

$$\min_{\mathbf{h}_2^{(n)}} \left(\mathbf{h}_2^{(n)}\right)^H \mathbf{\Gamma}_{\mathbf{h}_1^{(n)}}\mathbf{h}_2^{(n)} \quad \text{subject to} \quad \mathbf{C}_2^H\mathbf{h}_2^{(n)} = \mathbf{i}_\mathrm{c}, \quad (3.66)$$

where $\mathbf{\Gamma}_{\mathbf{h}_2^{(n-1)}}$ and $\mathbf{\Gamma}_{\mathbf{h}_1^{(n)}}$ are defined in (3.33) and (3.35), respectively. We easily get

$$\mathbf{h}_1^{(n)} = \mathbf{\Gamma}_{\mathbf{h}_2^{(n-1)}}^{-1}\mathbf{C}_1\left(\mathbf{C}_1^H\mathbf{\Gamma}_{\mathbf{h}_2^{(n-1)}}^{-1}\mathbf{C}_1\right)^{-1}\mathbf{i}_\mathrm{c} \quad (3.67)$$

and

$$\mathbf{h}_2^{(n)} = \mathbf{\Gamma}_{\mathbf{h}_1^{(n)}}^{-1}\mathbf{C}_2\left(\mathbf{C}_2^H\mathbf{\Gamma}_{\mathbf{h}_1^{(n)}}^{-1}\mathbf{C}_2\right)^{-1}\mathbf{i}_\mathrm{c}, \quad (3.68)$$

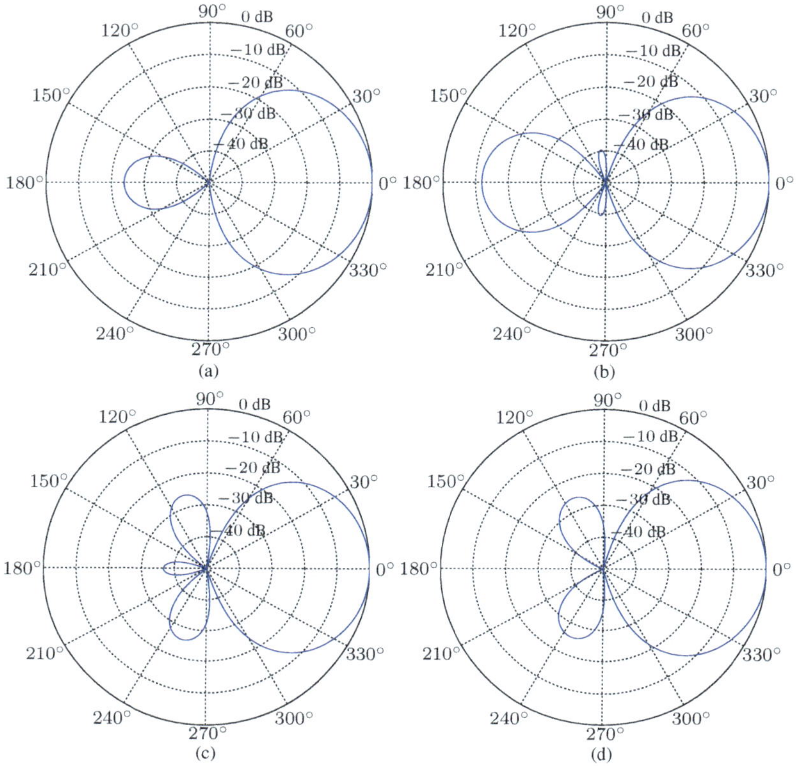

Fig. 3.11 Beampatterns of the sixth null-steering beamformer, $\mathbf{h}_{\mathrm{NS6}}$, for $\theta_{\mathrm{d}} = 0$, $f = 1$ kHz, $\delta = 5$ mm, $M_0 = 3$, and several values of θ_0: (a) $\theta_0 = 90°$, (b) $\theta_0 = 120°$, (c) $\theta_0 = 150°$, and (d) $\theta_0 = 180°$.

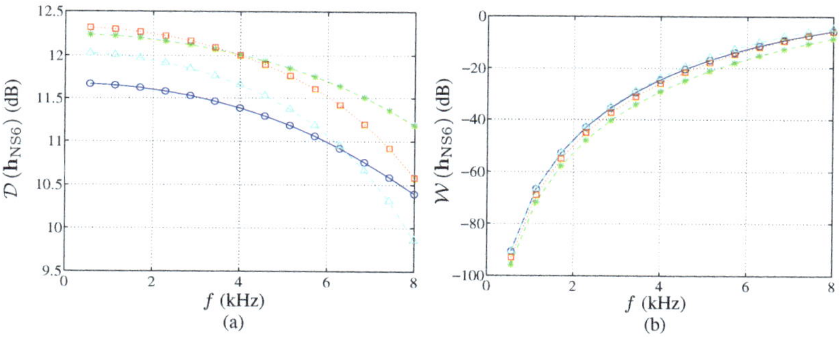

Fig. 3.12 Performance of the sixth null-steering beamformer, $\mathbf{h}_{\mathrm{NS6}}$, as a function of frequency for $\theta_{\mathrm{d}} = 0$, $\delta = 5$ mm, $M_0 = 3$, and several values of θ_0: $\theta_0 = 90°$ (solid line with circles), $\theta_0 = 120°$ (dashed line with asterisks), $\theta_0 = 150°$ (dotted line with squares), and $\theta_0 = 180°$ (dash-dot line with triangles). (a) DF and (b) WNG.

with the initialization:

$$\mathbf{h}_2^{(0)} = \mathbf{\Gamma}_2^{-1}\mathbf{C}_2\left(\mathbf{C}_2^H\mathbf{\Gamma}_2^{-1}\mathbf{C}_2\right)^{-1}\mathbf{i}_{\mathrm{c}}. \tag{3.69}$$

Therefore, the seventh and last proposed null-steering beamformer is at iteration n:

$$\mathbf{h}_{\mathrm{NS7}}^{(n)} = \mathbf{h}_1^{(n)} \otimes \mathbf{h}_2^{(n)}. \tag{3.70}$$

Figure 3.13 displays the directivity patterns of the seventh null-steering beamformer, $\mathbf{h}_{\mathrm{NS7}}^{(n)}$, for $f = 1$ kHz, $\delta = 5$ mm, $\theta_{\mathrm{d}} = 0$, $\theta_0 = 180°$, and $M_0 = 3$, obtained at the iteration n for several values of n. Figure 3.14 shows plots of the DFs and WNGs of the seventh null-steering beamformer as a function of frequency for $\delta = 5$ mm, $\theta_{\mathrm{d}} = 0$, $\theta_0 = 180°$, $M_0 = 3$, and several values of n. We observe that the DF of the seventh null-steering beamformer increases at each iteration, and roughly converges after two iterations, while the WNG decreases at each iteration. Compared with the above null-steering beamformers, the seventh null-steering beamformer yields the highest DF, but the lowest WNG (compare Figs 3.12 and 3.14).

3.2 Adaptive Beamformers

Before developing some useful adaptive beamformers in our context, we first present other important performance measures that depend on the second-order statistics of the signals.

3.2.1 Other Measures

The (narrowband) noise reduction factor quantifies the amount of noise being rejected by the beamformer. It is defined as

$$\xi_{\mathrm{n}}(\mathbf{h}) = \frac{\phi_{V_1}}{\mathbf{h}^H\mathbf{\Phi}_{\mathbf{v}}\mathbf{h}}. \tag{3.71}$$

The noise reduction factor is expected to be lower bounded by 1; otherwise, the beamformer amplifies the noise. The higher the value of the noise reduction factor, the more the noise is rejected. While the output SNR is upper bounded, the noise reduction factor is not. In the distortionless case, i.e., $\mathbf{h}^H\mathbf{d}_{\theta_{\mathrm{d}}} = 1$, the noise reduction factor coincides with the array gain [defined in (2.26)].

Since the noise is reduced by the beamforming operation, so is, in general, the desired signal. This desired signal reduction (or cancellation) implies, in

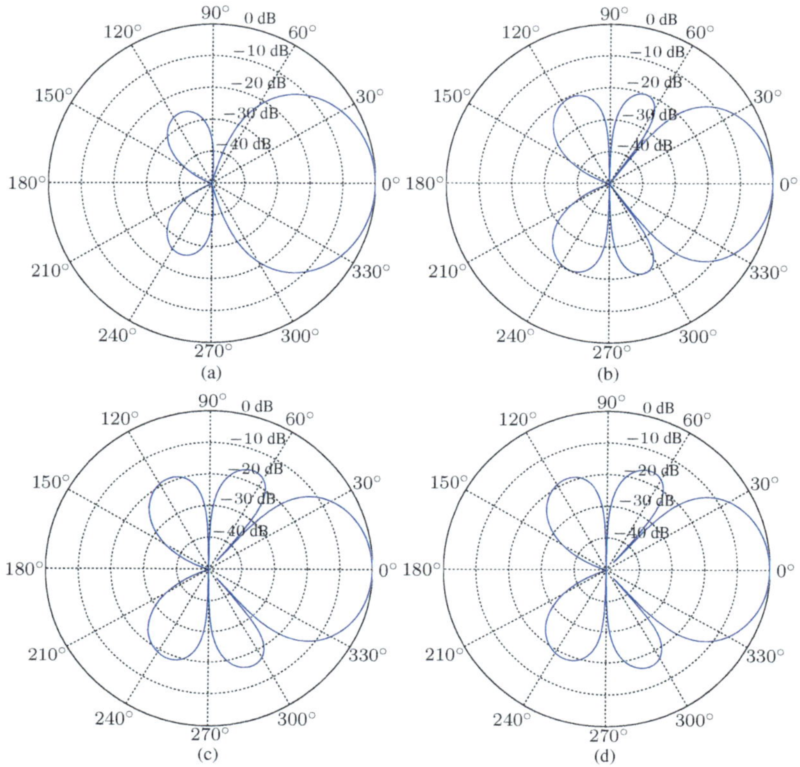

Fig. 3.13 Beampatterns of the seventh null-steering beamformer, $\mathbf{h}_{NS7}^{(n)}$, for $f = 1$ kHz, $\delta = 5$ mm, $\theta_d = 0$, $\theta_0 = 180°$, and $M_0 = 3$, obtained at the iteration n: (a) $n = 0$, (b) $n = 1$, (c) $n = 2$, and (d) $n = 10$.

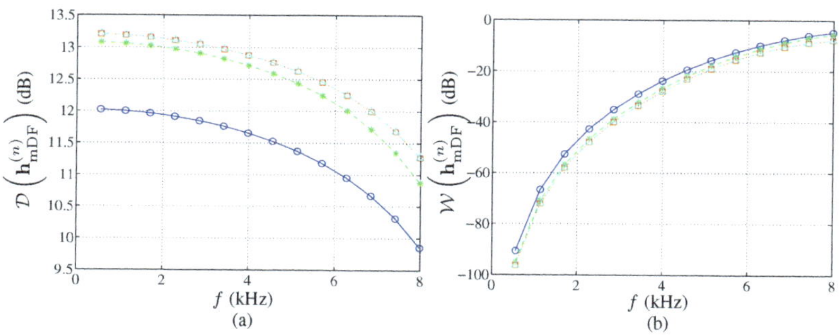

Fig. 3.14 Performance of the seventh null-steering beamformer, $\mathbf{h}_{NS7}^{(n)}$, as a function of frequency for $\delta = 5$ mm, $\theta_d = 0$, $\theta_0 = 180°$, $M_0 = 3$, and several values of n: $n = 0$ (solid line with circles), $n = 1$ (dashed line with asterisks), $n = 2$ (dotted line with squares), and $n = 10$ (dash-dot line with triangles). (a) DF and (b) WNG.

general, distortion. The (narrowband) desired signal reduction factor is

$$\begin{aligned}\xi_{\mathrm{d}}(\mathbf{h}) &= \frac{\phi_X}{\mathbf{h}^H \mathbf{\Phi}_{\mathbf{x}} \mathbf{h}} \\ &= \frac{1}{\left|\mathbf{h}^H \mathbf{d}_{\theta_{\mathrm{d}}}\right|^2} \\ &= \frac{1}{\left|\mathbf{h}_1^H \mathbf{d}_{1,\theta_{\mathrm{d}}}\right|^2} \times \frac{1}{\left|\mathbf{h}_2^H \mathbf{d}_{2,\theta_{\mathrm{d}}}\right|^2} \\ &= \xi_{1,\mathrm{d}}(\mathbf{h}_1) \times \xi_{2,\mathrm{d}}(\mathbf{h}_2).\end{aligned} \tag{3.72}$$

The closer the value of $\xi_{\mathrm{d}}(\mathbf{h})$ is to 1, the less distorted is the desired signal.

It is easy to verify that we have the following fundamental relation:

$$\frac{\mathrm{oSNR}(\mathbf{h})}{\mathrm{iSNR}} = \frac{\xi_{\mathrm{n}}(\mathbf{h})}{\xi_{\mathrm{d}}(\mathbf{h})}, \tag{3.73}$$

where the output and input SNRs are defined in (2.25) and (2.24), respectively. This expression indicates the equivalence between gain/loss in SNR and distortion (of both the desired and noise signals).

Another way to measure the distortion of the desired signal due to the beamforming operation is via the (narrowband) desired signal distortion index:

$$\begin{aligned}\upsilon_{\mathrm{d}}(\mathbf{h}) &= \frac{E\left(\left|X_{\mathrm{fd}} - X\right|^2\right)}{\phi_X} \\ &= \left|\mathbf{h}^H \mathbf{d}_{\theta_{\mathrm{d}}} - 1\right|^2.\end{aligned} \tag{3.74}$$

The desired signal distortion index is close to 0 if there is no distortion and expected to be greater than 0 when distortion occurs.

Error criteria play a critical role in deriving optimal beamformers. The mean-squared error (MSE) [3] is, by far, the most practical one. We define the error signal between the estimated and desired signals as

$$\begin{aligned}\mathcal{E} &= Z - X \\ &= X_{\mathrm{fd}} + V_{\mathrm{rn}} - X \\ &= \mathcal{E}_{\mathrm{d}} + \mathcal{E}_{\mathrm{n}},\end{aligned} \tag{3.75}$$

where

$$\mathcal{E}_{\mathrm{d}} = \left(\mathbf{h}^H \mathbf{d}_{\theta_{\mathrm{d}}} - 1\right) X \tag{3.76}$$

is the desired signal distortion due to the beamformer and

$$\mathcal{E}_{\mathrm{n}} = \mathbf{h}^H \mathbf{v} \tag{3.77}$$

represents the residual noise. Since $\mathcal{E}_\mathrm{d}$ and $\mathcal{E}_\mathrm{n}$ are incoherent, the (narrowband) MSE can be expressed as

$$\begin{aligned} J(\mathbf{h}) &= E\left(|\mathcal{E}|^2\right) \\ &= E\left(|\mathcal{E}_\mathrm{d}|^2\right) + E\left(|\mathcal{E}_\mathrm{n}|^2\right) \\ &= J_\mathrm{d}(\mathbf{h}) + J_\mathrm{n}(\mathbf{h}) \\ &= \phi_X + \mathbf{h}^H \mathbf{\Phi}_\mathbf{y} \mathbf{h} - \phi_X \mathbf{h}^H \mathbf{d}_{\theta_\mathrm{d}} - \phi_X \mathbf{d}_{\theta_\mathrm{d}}^H \mathbf{h}, \end{aligned} \tag{3.78}$$

where

$$\begin{aligned} J_\mathrm{d}(\mathbf{h}) &= \phi_X \left|\mathbf{h}^H \mathbf{d}_{\theta_\mathrm{d}} - 1\right|^2 \\ &= \phi_X \upsilon_\mathrm{d}(\mathbf{h}) \end{aligned} \tag{3.79}$$

and

$$\begin{aligned} J_\mathrm{n}(\mathbf{h}) &= \mathbf{h}^H \mathbf{\Phi}_\mathbf{v} \mathbf{h} \\ &= \frac{\phi_{V_1}}{\xi_\mathrm{n}(\mathbf{h})}. \end{aligned} \tag{3.80}$$

We have the following classical relationships:

$$\begin{aligned} \frac{J_\mathrm{d}(\mathbf{h})}{J_\mathrm{n}(\mathbf{h})} &= \mathrm{iSNR} \times \xi_\mathrm{n}(\mathbf{h}) \times \upsilon_\mathrm{d}(\mathbf{h}) \\ &= \mathrm{oSNR}(\mathbf{h}) \times \xi_\mathrm{d}(\mathbf{h}) \times \upsilon_\mathrm{d}(\mathbf{h}). \end{aligned} \tag{3.81}$$

3.2.2 Wiener

Because of the structure of $\mathbf{h}$, a closed-form Wiener beamformer cannot apparently be found but an iterative one can be derived. Using (2.40) and (2.41), we can express the MSE in (3.78) as

$$J(\mathbf{h}_1 \otimes \mathbf{h}_2) = \phi_X + \mathbf{h}_1^H \mathbf{\Phi}_{\mathbf{y},2} \mathbf{h}_1 - \phi_{X,2} \mathbf{h}_1^H \mathbf{d}_{1,\theta_\mathrm{d}} - \phi_{X,2}^* \mathbf{d}_{1,\theta_\mathrm{d}}^H \mathbf{h}_1 \tag{3.82}$$

$$= \phi_X + \mathbf{h}_2^H \mathbf{\Phi}_{\mathbf{y},1} \mathbf{h}_2 - \phi_{X,1} \mathbf{h}_2^H \mathbf{d}_{2,\theta_\mathrm{d}} - \phi_{X,1}^* \mathbf{d}_{2,\theta_\mathrm{d}}^H \mathbf{h}_2, \tag{3.83}$$

where

$$\mathbf{\Phi}_{\mathbf{y},2} = \left(\mathbf{I}_{M_0} \otimes \mathbf{h}_2\right)^H \mathbf{\Phi}_\mathbf{y} \left(\mathbf{I}_{M_0} \otimes \mathbf{h}_2\right), \tag{3.84}$$

$$\phi_{X,2} = \phi_X \mathbf{h}_2^H \mathbf{d}_{2,\theta_\mathrm{d}}, \tag{3.85}$$

and

$$\mathbf{\Phi}_{\mathbf{y},1} = (\mathbf{h}_1 \otimes \mathbf{I}_{M_0})^H \mathbf{\Phi}_{\mathbf{y}} (\mathbf{h}_1 \otimes \mathbf{I}_{M_0}), \tag{3.86}$$

$$\phi_{X,1} = \phi_X \mathbf{h}_1^H \mathbf{d}_{1,\theta_\mathrm{d}}. \tag{3.87}$$

It is interesting to notice that the size of the matrices $\mathbf{\Phi}_{\mathbf{y},1}$ and $\mathbf{\Phi}_{\mathbf{y},2}$, which is $M_0 \times M_0$, is much smaller than the size of $\mathbf{\Phi}_{\mathbf{y}}$, which is $M_0^2 \times M_0^2$. As a result, in practice, much less observations are needed to accurately estimate $\mathbf{\Phi}_{\mathbf{y},1}$ and $\mathbf{\Phi}_{\mathbf{y},2}$ than $\mathbf{\Phi}_{\mathbf{y}}$, which is the matrix that is inverted in the conventional Wiener beamformer.

When $\mathbf{h}_2$ is fixed, we write (3.82) as

$$J(\mathbf{h}_1|\mathbf{h}_2) = \phi_X + \mathbf{h}_1^H \mathbf{\Phi}_{\mathbf{y},2}\mathbf{h}_1 - \phi_{X,2}\mathbf{h}_1^H \mathbf{d}_{1,\theta_\mathrm{d}} - \phi_{X,2}^* \mathbf{d}_{1,\theta_\mathrm{d}}^H \mathbf{h}_1, \tag{3.88}$$

and when $\mathbf{h}_1$ is fixed, we write (3.83) as

$$J(\mathbf{h}_2|\mathbf{h}_1) = \phi_X + \mathbf{h}_2^H \mathbf{\Phi}_{\mathbf{y},1}\mathbf{h}_2 - \phi_{X,1}\mathbf{h}_2^H \mathbf{d}_{2,\theta_\mathrm{d}} - \phi_{X,1}^* \mathbf{d}_{2,\theta_\mathrm{d}}^H \mathbf{h}_2. \tag{3.89}$$

Now, we have everything to derive an iterative algorithm similar to the one proposed in [4]. At iteration 0, we may take

$$\mathbf{h}_2^{(0)} = \phi_X \mathbf{\Phi}_{\mathbf{y}_2}^{-1} \mathbf{d}_{2,\theta_\mathrm{d}}, \tag{3.90}$$

where $\mathbf{\Phi}_{\mathbf{y}_2}$ is the covariance matrix of

$$\mathbf{y}_2 = \begin{bmatrix} Y_1 \ Y_2 \ \cdots \ Y_{M_0} \end{bmatrix}^T, \tag{3.91}$$

whose elements are the M_0 first ones of $\mathbf{y}$. In fact, $\mathbf{h}_2^{(0)}$ is just the traditional Wiener beamformer applied to the second ULA. Substituting $\mathbf{h}_2^{(0)}$ into (3.84)–(3.85), we get

$$\mathbf{\Phi}_{\mathbf{y},2}^{(0)} = \left(\mathbf{I}_{M_0} \otimes \mathbf{h}_2^{(0)}\right)^H \mathbf{\Phi}_{\mathbf{y}} \left(\mathbf{I}_{M_0} \otimes \mathbf{h}_2^{(0)}\right), \tag{3.92}$$

$$\phi_{X,2}^{(0)} = \phi_X \left(\mathbf{h}_2^{(0)}\right)^H \mathbf{d}_{2,\theta_\mathrm{d}}. \tag{3.93}$$

Then, substituting these quantities into the MSE in (3.88), we obtain at iteration 1:

$$\begin{aligned} J\left(\mathbf{h}_1^{(1)}|\mathbf{h}_2^{(0)}\right) &= \phi_X + \left(\mathbf{h}_1^{(1)}\right)^H \mathbf{\Phi}_{\mathbf{y},2}^{(0)}\mathbf{h}_1^{(1)} - \phi_{X,2}^{(0)} \left(\mathbf{h}_1^{(1)}\right)^H \mathbf{d}_{1,\theta_\mathrm{d}} \\ &\quad - \left(\phi_{X,2}^{(0)}\right)^* \mathbf{d}_{1,\theta_\mathrm{d}}^H \mathbf{h}_1^{(1)}. \end{aligned} \tag{3.94}$$

The minimization of $J\left(\mathbf{h}_1^{(1)}|\mathbf{h}_2^{(0)}\right)$ with respect to $\mathbf{h}_1^{(1)}$ gives

$$\mathbf{h}_1^{(1)} = \phi_{X,2}^{(0)} \left(\mathbf{\Phi}_{\mathbf{y},2}^{(0)}\right)^{-1} \mathbf{d}_{1,\theta_\mathrm{d}}. \tag{3.95}$$

Using $\mathbf{h}_1^{(1)}$ into (3.86)–(3.87), we obtain

$$\mathbf{\Phi}_{\mathbf{y},1}^{(1)} = \left(\mathbf{h}_1^{(1)} \otimes \mathbf{I}_{M_0}\right)^H \mathbf{\Phi}_{\mathbf{y}} \left(\mathbf{h}_1^{(1)} \otimes \mathbf{I}_{M_0}\right), \tag{3.96}$$

$$\phi_{X,1}^{(1)} = \phi_X \left(\mathbf{h}_1^{(1)}\right)^H \mathbf{d}_{1,\theta_{\mathrm{d}}}. \tag{3.97}$$

With $\mathbf{\Phi}_{\mathbf{y},1}^{(1)}$ and $\phi_{X,1}^{(1)}$, we can compute the MSE in (3.89) as

$$J\left(\mathbf{h}_2^{(1)}|\mathbf{h}_1^{(1)}\right) = \phi_X + \left(\mathbf{h}_2^{(1)}\right)^H \mathbf{\Phi}_{\mathbf{y},1}^{(1)}\mathbf{h}_2^{(1)} - \phi_{X,1}^{(1)}\left(\mathbf{h}_2^{(1)}\right)^H \mathbf{d}_{2,\theta_{\mathrm{d}}} - \left(\phi_{X,1}^{(1)}\right)^* \mathbf{d}_{2,\theta_{\mathrm{d}}}^H \mathbf{h}_2^{(1)}, \tag{3.98}$$

whose minimization with respect to $\mathbf{h}_2^{(1)}$ gives

$$\mathbf{h}_2^{(1)} = \phi_{X,1}^{(1)} \left(\mathbf{\Phi}_{\mathbf{y},1}^{(1)}\right)^{-1} \mathbf{d}_{2,\theta_{\mathrm{d}}}. \tag{3.99}$$

Continuing the iterations up to the iteration n, we easily get the estimate of the first beamformer:

$$\mathbf{h}_1^{(n)} = \phi_{X,2}^{(n-1)} \left(\mathbf{\Phi}_{\mathbf{y},2}^{(n-1)}\right)^{-1} \mathbf{d}_{1,\theta_{\mathrm{d}}}, \tag{3.100}$$

where

$$\phi_{X,2}^{(n-1)} = \phi_X \left(\mathbf{h}_2^{(n-1)}\right)^H \mathbf{d}_{2,\theta_{\mathrm{d}}}, \tag{3.101}$$

$$\mathbf{\Phi}_{\mathbf{y},2}^{(n-1)} = \left(\mathbf{I}_{M_0} \otimes \mathbf{h}_2^{(n-1)}\right)^H \mathbf{\Phi}_{\mathbf{y}} \left(\mathbf{I}_{M_0} \otimes \mathbf{h}_2^{(n-1)}\right), \tag{3.102}$$

and the estimate of the second beamformer:

$$\mathbf{h}_2^{(n)} = \phi_{X,1}^{(n)} \left(\mathbf{\Phi}_{\mathbf{y},1}^{(n)}\right)^{-1} \mathbf{d}_{2,\theta_{\mathrm{d}}}, \tag{3.103}$$

where

$$\phi_{X,1}^{(n)} = \phi_X \left(\mathbf{h}_1^{(n)}\right)^H \mathbf{d}_{1,\theta_{\mathrm{d}}}, \tag{3.104}$$

$$\mathbf{\Phi}_{\mathbf{y},1}^{(n)} = \left(\mathbf{h}_1^{(n)} \otimes \mathbf{I}_{M_0}\right)^H \mathbf{\Phi}_{\mathbf{y}} \left(\mathbf{h}_1^{(n)} \otimes \mathbf{I}_{M_0}\right). \tag{3.105}$$

Finally, we deduce that the Wiener beamformer is at iteration n:

$$\mathbf{h}_{\mathrm{W}}^{(n)} = \mathbf{h}_1^{(n)} \otimes \mathbf{h}_2^{(n)}, \tag{3.106}$$

where $\mathbf{h}_1^{(n)}$ and $\mathbf{h}_2^{(n)}$ are defined in (3.100) and (3.103), respectively.

Example 3.1. Suppose that a desired signal impinges on the ULA from the direction θ_{d}, and that a statistically independent interference impinges on the ULA from the direction θ_0. Assume that the desired signal is a harmonic pulse of T samples:

$$x(t) = \begin{cases} A \sin(\omega_0 t + \phi), & 0 \leq t \leq T-1 \\ 0, & t < 0, \ t \geq T \end{cases},$$

with fixed amplitude A and angular frequency ω_0, and random phase ϕ, uniformly distributed on the interval from 0 to 2π. Assume that the interference $u(t)$ is white Gaussian noise, i.e., $u(t) \sim \mathcal{N}(0, \sigma_u^2)$, uncorrelated with $x(t)$. In addition, the sensors contain thermal white Gaussian noise, $w_m(t) \sim \mathcal{N}(0, \sigma_w^2)$, that are mutually uncorrelated. The noisy received signals are given by $y_m(t) = x_m(t) + v_m(t)$, $m = 1, 2, \ldots, M$, where $v_m(t) = u_m(t) + w_m(t)$, $m = 1, 2, \ldots, M$ are the interference-plus-noise signals. The variance of $X(\omega)$ is given by

$$\phi_X = \frac{A^2}{4} D_T^2 [\pi(\omega + \omega_0)] + \frac{A^2}{4} D_T^2 [\pi(\omega - \omega_0)],$$

where

$$D_T(x) = \frac{\sin(Tx)}{\sin(x)}.$$

The covariance matrices of $\mathbf{x}(\omega)$ and $\mathbf{v}(\omega)$ are given by

$$\begin{aligned} \mathbf{\Phi}_{\mathbf{x}} &= \phi_X \mathbf{d}_{\theta_{\mathrm{d}}} \mathbf{d}_{\theta_{\mathrm{d}}}^H, \\ \mathbf{\Phi}_{\mathbf{v}} &= T\sigma_u^2 \mathbf{d}_{\theta_0} \mathbf{d}_{\theta_0}^H + T\sigma_w^2 \mathbf{I}_M. \end{aligned}$$

To demonstrate the performance of the Wiener beamformer, we choose $A = 0.5$, $\omega_0 = 2\pi f_0$, $f_0 = 3$ kH, $T = 500$, $\theta_{\mathrm{d}} = 70°$, $\theta_0 = 30°$, and $\sigma_w^2 = 0.01\sigma_u^2$. Figure 3.15 displays the directivity patterns of the Wiener beamformer, $\mathbf{h}_{\mathrm{W}}^{(n)}$, for iSNR $= 0$ dB, $f = 3$ kHz, $\delta = 1$ cm, and $M_0 = 5$, obtained at the iteration n for several values of n. The directivity patterns converge after three iterations. The main beam is in the direction of the desired signal, i.e., θ_{d}, and there is a null in the direction of the interference, i.e., θ_0. Figure 3.16 shows plots of the gain in SNR, $\mathcal{G}\left(\mathbf{h}_{\mathrm{W}}^{(n)}\right)$, the noise reduction factor, $\xi_{\mathrm{n}}\left(\mathbf{h}_{\mathrm{W}}^{(n)}\right)$, the desired signal distortion index, $\upsilon_{\mathrm{d}}\left(\mathbf{h}_{\mathrm{W}}^{(n)}\right)$, and the MSE, $J\left(\mathbf{h}_{\mathrm{W}}^{(n)}\right)$, as a function of the input SNR for $f = 3$ kHz, $\delta = 1$ cm, $M_0 = 5$, and several values of n. We observe that the MSE and the desired signal distortion index obtained by the Wiener beamformer decrease at each iteration, and roughly converge after three iterations, while the gain in SNR and the noise reduction factor increase at each iteration. Figure 3.17 displays the directivity patterns of the Wiener beamformer at the iteration $n = 10$, for

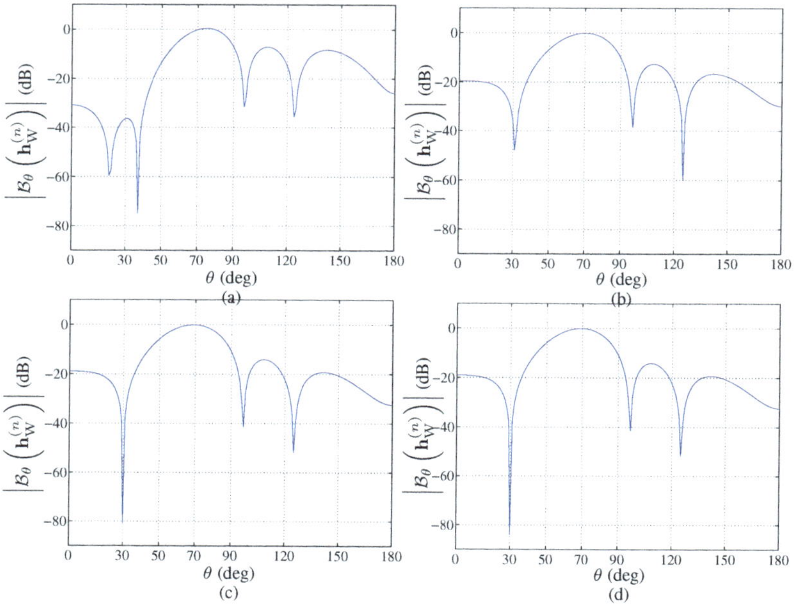

Fig. 3.15 Beampatterns of the Wiener beamformer, $\mathbf{h}_{\mathrm{W}}^{(n)}$, for iSNR = 0 dB, $f = 3$ kHz, $\delta = 1$ cm, $\theta_{\mathrm{d}} = 70°$, $\theta_0 = 30°$, and $M_0 = 5$, obtained at the iteration n: (a) $n = 1$, (b) $n = 2$, (c) $n = 3$, and (d) $n = 4$.

iSNR = 0 dB, $f = 3$ kHz, $\delta = 1$ cm, and different numbers of sensors, M. As the number of sensors increases, the width of the main beam decreases, and the null in the direction of the interference becomes deeper. Figure 3.18 shows plots of the gain in SNR, $\mathcal{G}\left(\mathbf{h}_{\mathrm{W}}^{(10)}\right)$, the noise reduction factor, $\xi_{\mathrm{n}}\left(\mathbf{h}_{\mathrm{W}}^{(10)}\right)$, the desired signal distortion index, $\upsilon_{\mathrm{d}}\left(\mathbf{h}_{\mathrm{W}}^{(10)}\right)$, and the MSE, $J\left(\mathbf{h}_{\mathrm{W}}^{(10)}\right)$, as a function of the input SNR for $f = 3$ kHz, $\delta = 1$ cm, and different numbers of sensors, M. We observe that as the number of sensors increases, the MSE and the desired signal distortion index obtained by the Wiener beamformer decrease while the gain in SNR and the noise reduction factor increase.

3.2.3 Tradeoff

In order to better compromise between noise reduction and desired signal distortion, we can minimize the desired signal distortion indices with the constraints that the noise reduction factors are equal to positive values that are greater than 1, i.e.,

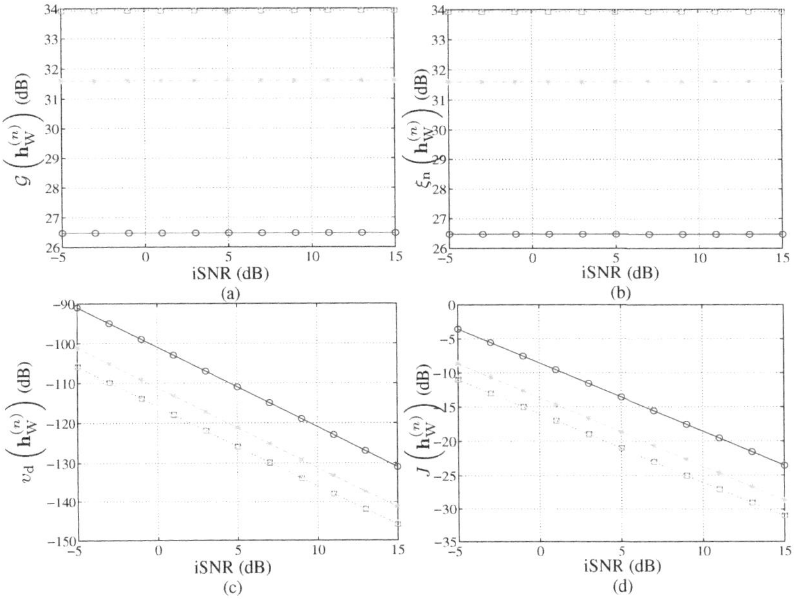

Fig. 3.16 Performance of the Wiener beamformer, $\mathbf{h}_{\mathrm{W}}^{(n)}$, as a function of the input SNR for $f = 3$ kHz, $\delta = 1$ cm, $\theta_{\mathrm{d}} = 70°$, $\theta_0 = 30°$, $M_0 = 5$, and several values of n: $n = 1$ (solid line with circles), $n = 2$ (dashed line with asterisks), $n = 3$ (dotted line with squares), and $n = 10$ (dash-dot line with triangles). (a) Gain in SNR, (b) noise reduction factor, (c) desired signal distortion index, and (d) MSE.

$$\min_{\mathbf{h}_1^{(n)}} J_{\mathrm{d}}\left(\mathbf{h}_1^{(n)}|\mathbf{h}_2^{(n-1)}\right) \quad \text{subject to} \quad J_{\mathrm{n}}\left(\mathbf{h}_1^{(n)}|\mathbf{h}_2^{(n-1)}\right) = \aleph_1 \phi_{V_1}, \tag{3.107}$$

$$\min_{\mathbf{h}_2^{(n)}} J_{\mathrm{d}}\left(\mathbf{h}_2^{(n)}|\mathbf{h}_1^{(n)}\right) \quad \text{subject to} \quad J_{\mathrm{n}}\left(\mathbf{h}_2^{(n)}|\mathbf{h}_1^{(n)}\right) = \aleph_2 \phi_{V_1}, \tag{3.108}$$

where $0 < \aleph_1, \aleph_2 < 1$ to insure that we get some noise reduction,

$$J_{\mathrm{d}}\left(\mathbf{h}_1^{(n)}|\mathbf{h}_2^{(n-1)}\right) = \phi_X + \left(\mathbf{h}_1^{(n)}\right)^H \mathbf{\Phi}_{\mathbf{x},2}^{(n-1)} \mathbf{h}_1^{(n)} - \phi_{X,2}^{(n-1)} \left(\mathbf{h}_1^{(n)}\right)^H \mathbf{d}_{1,\theta_{\mathrm{d}}} - \left(\phi_{X,2}^{(n-1)}\right)^* \mathbf{d}_{1,\theta_{\mathrm{d}}}^H \mathbf{h}_1^{(n)}, \tag{3.109}$$

$$J_{\mathrm{n}}\left(\mathbf{h}_1^{(n)}|\mathbf{h}_2^{(n-1)}\right) = \left(\mathbf{h}_1^{(n)}\right)^H \mathbf{\Phi}_{\mathbf{v},2}^{(n-1)} \mathbf{h}_1^{(n)}, \tag{3.110}$$

and

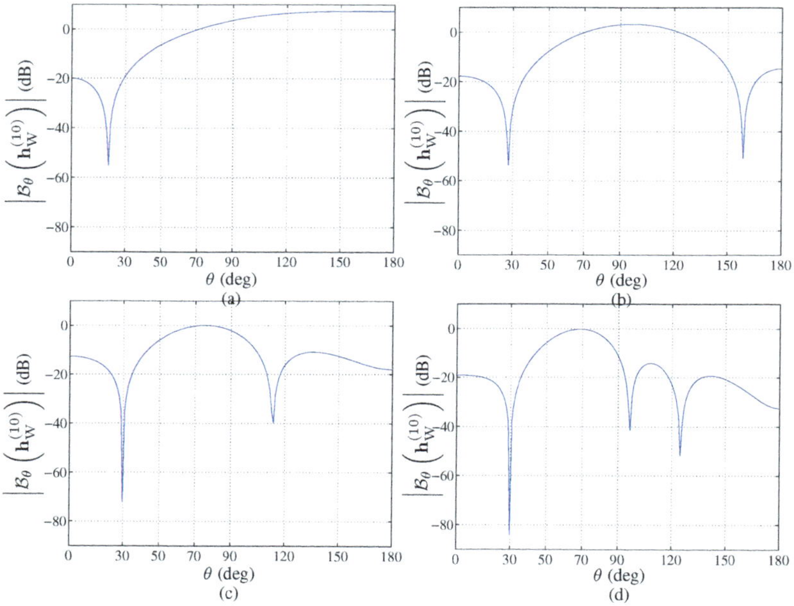

Fig. 3.17 Beampatterns of the Wiener beamformer at the iteration $n = 10$, $\mathbf{h}_W^{(10)}$, for iSNR = 0 dB, $f = 3$ kHz, $\delta = 1$ cm, $\theta_d = 70°$, $\theta_0 = 30°$, and different numbers of sensors $M = M_0^2$: (a) $M_0 = 2$, (b) $M_0 = 3$, (c) $M_0 = 4$, and (d) $M_0 = 5$.

$$J_d\left(\mathbf{h}_2^{(n)}|\mathbf{h}_1^{(n)}\right) = \phi_X + \left(\mathbf{h}_2^{(n)}\right)^H \mathbf{\Phi}_{\mathbf{x},1}^{(n)}\mathbf{h}_2^{(n)} - \phi_{X,1}^{(n)}\left(\mathbf{h}_2^{(n)}\right)^H \mathbf{d}_{2,\theta_d} - \left(\phi_{X,1}^{(n)}\right)^* \mathbf{d}_{2,\theta_d}^H \mathbf{h}_2^{(n)}, \quad (3.111)$$

$$J_n\left(\mathbf{h}_2^{(n)}|\mathbf{h}_1^{(n)}\right) = \left(\mathbf{h}_2^{(n)}\right)^H \mathbf{\Phi}_{\mathbf{v},1}^{(n)}\mathbf{h}_2^{(n)}, \quad (3.112)$$

with

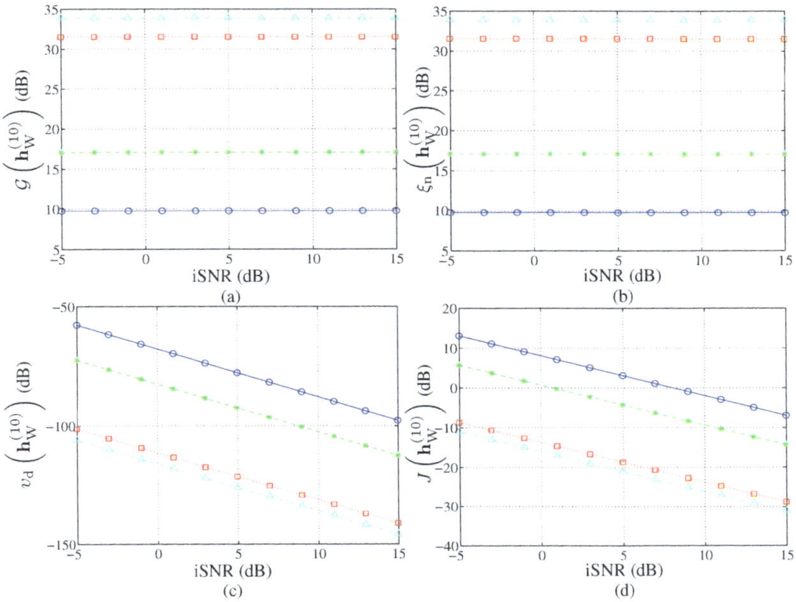

Fig. 3.18 Performance of the Wiener beamformer at the iteration $n = 10$, $\mathbf{h}_{\mathrm{W}}^{(10)}$, as a function of the input SNR for $f = 3$ kHz, $\delta = 1$ cm, $\theta_{\mathrm{d}} = 70°$, $\theta_0 = 30°$, and different numbers of sensors $M = M_0^2$: $M_0 = 2$ (solid line with circles), $M_0 = 3$ (dashed line with asterisks), $M_0 = 4$ (dotted line with squares), and $M_0 = 5$ (dash-dot line with triangles). (a) Gain in SNR, (b) noise reduction factor, (c) desired signal distortion index, and (d) MSE.

$$\mathbf{\Phi}_{\mathbf{x},2}^{(n-1)} = \left(\mathbf{I}_{M_0} \otimes \mathbf{h}_2^{(n-1)}\right)^H \mathbf{\Phi}_{\mathbf{x}} \left(\mathbf{I}_{M_0} \otimes \mathbf{h}_2^{(n-1)}\right) = \frac{\left|\phi_{X,2}^{(n-1)}\right|^2}{\phi_X} \mathbf{d}_{1,\theta_{\mathrm{d}}} \mathbf{d}_{1,\theta_{\mathrm{d}}}^H, \tag{3.113}$$

$$\mathbf{\Phi}_{\mathbf{v},2}^{(n-1)} = \left(\mathbf{I}_{M_0} \otimes \mathbf{h}_2^{(n-1)}\right)^H \mathbf{\Phi}_{\mathbf{v}} \left(\mathbf{I}_{M_0} \otimes \mathbf{h}_2^{(n-1)}\right), \tag{3.114}$$

$$\mathbf{\Phi}_{\mathbf{x},1}^{(n)} = \left(\mathbf{h}_1^{(n)} \otimes \mathbf{I}_{M_0}\right)^H \mathbf{\Phi}_{\mathbf{x}} \left(\mathbf{h}_1^{(n)} \otimes \mathbf{I}_{M_0}\right) = \frac{\left|\phi_{X,1}^{(n)}\right|^2}{\phi_X} \mathbf{d}_{2,\theta_{\mathrm{d}}} \mathbf{d}_{2,\theta_{\mathrm{d}}}^H, \tag{3.115}$$

$$\mathbf{\Phi}_{\mathbf{v},1}^{(n)} = \left(\mathbf{h}_1^{(n)} \otimes \mathbf{I}_{M_0}\right)^H \mathbf{\Phi}_{\mathbf{v}} \left(\mathbf{h}_1^{(n)} \otimes \mathbf{I}_{M_0}\right). \tag{3.116}$$

By using Lagrange multipliers, $\mu_1, \mu_2 > 0$, to adjoin the constraints to the cost functions, we get

$$\mathbf{h}_{1,\mu_1}^{(n)} = \phi_{X,2}^{(n-1)} \left(\mathbf{\Phi}_{\mathbf{x},2}^{(n-1)} + \mu_1 \mathbf{\Phi}_{\mathbf{v},2}^{(n-1)} \right)^{-1} \mathbf{d}_{1,\theta_\mathrm{d}} \\ = \frac{\phi_{X,2}^{(n-1)} \phi_X \left(\mathbf{\Phi}_{\mathbf{v},2}^{(n-1)} \right)^{-1} \mathbf{d}_{1,\theta_\mathrm{d}}}{\mu_1 \phi_X + \left| \phi_{X,2}^{(n-1)} \right|^2 \mathbf{d}_{1,\theta_\mathrm{d}}^H \left(\mathbf{\Phi}_{\mathbf{v},2}^{(n-1)} \right)^{-1} \mathbf{d}_{1,\theta_\mathrm{d}}} \tag{3.117}$$

and

$$\mathbf{h}_{2,\mu_2}^{(n)} = \phi_{X,1}^{(n)} \left(\mathbf{\Phi}_{\mathbf{x},1}^{(n)} + \mu_2 \mathbf{\Phi}_{\mathbf{v},1}^{(n)} \right)^{-1} \mathbf{d}_{2,\theta_\mathrm{d}} \\ = \frac{\phi_{X,1}^{(n)} \phi_X \left(\mathbf{\Phi}_{\mathbf{v},1}^{(n)} \right)^{-1} \mathbf{d}_{2,\theta_\mathrm{d}}}{\mu_2 \phi_X + \left| \phi_{X,1}^{(n)} \right|^2 \mathbf{d}_{2,\theta_\mathrm{d}}^H \left(\mathbf{\Phi}_{\mathbf{v},1}^{(n)} \right)^{-1} \mathbf{d}_{2,\theta_\mathrm{d}}}, \tag{3.118}$$

with the initialization:

$$\mathbf{h}_{2,\mu_2}^{(0)} = \phi_X \left(\phi_X \mathbf{d}_{2,\theta_\mathrm{d}} \mathbf{d}_{2,\theta_\mathrm{d}}^H + \mu_2 \mathbf{\Phi}_{\mathbf{v}_2} \right)^{-1} \mathbf{d}_{2,\theta_\mathrm{d}} \\ = \frac{\phi_X \mathbf{\Phi}_{\mathbf{v}_2}^{-1} \mathbf{d}_{2,\theta_\mathrm{d}}}{\mu_2 + \phi_X \mathbf{d}_{2,\theta_\mathrm{d}}^H \mathbf{\Phi}_{\mathbf{v}_2}^{-1} \mathbf{d}_{2,\theta_\mathrm{d}}}, \tag{3.119}$$

where $\mathbf{\Phi}_{\mathbf{v}_2}$ is the covariance matrix of

$$\mathbf{v}_2 = \begin{bmatrix} V_1 \; V_2 \; \cdots \; V_{M_0} \end{bmatrix}^T, \tag{3.120}$$

whose elements are the M_0 first ones of $\mathbf{v}$. As a matter of fact, $\mathbf{h}_{2,\mu_2}^{(0)}$ is just the traditional tradeoff beamformer applied to the second ULA. Therefore, we find that the tradeoff beamformer is at iteration n:

$$\mathbf{h}_{\mathrm{T},\mu_1,\mu_2}^{(n)} = \mathbf{h}_{1,\mu_1}^{(n)} \otimes \mathbf{h}_{2,\mu_2}^{(n)}, \tag{3.121}$$

where $\mathbf{h}_{1,\mu_1}^{(n)}$ and $\mathbf{h}_{2,\mu_2}^{(n)}$ are defined in (3.117) and (3.118), respectively. We can see that for

- $\mu_1 = \mu_2 = 1$, we get the Wiener beamformer;
- $\mu_1, \mu_2 > 1$, results in a beamformer with low residual noise at the expense of high desired signal distortion (as compared to Wiener); and
- $\mu_1, \mu_2 < 1$, results in a beamformer with high residual noise and low desired signal distortion (as compared to Wiener).

3.2.4 MVDR

The minimum variance distortionless response (MVDR) beamformer proposed by Capon [5], [6] is obtained by minimizing the MSEs of the residual noise subject to the distortionless constraints, i.e.,

$$\min_{\mathbf{h}_1^{(n)}} \left(\mathbf{h}_1^{(n)}\right)^H \mathbf{\Phi}_{\mathbf{v},2}^{(n-1)} \mathbf{h}_1^{(n)} \quad \text{subject to} \quad \left(\mathbf{h}_1^{(n)}\right)^H \mathbf{d}_{1,\theta_\mathrm{d}} = 1, \tag{3.122}$$

$$\min_{\mathbf{h}_2^{(n)}} \left(\mathbf{h}_2^{(n)}\right)^H \mathbf{\Phi}_{\mathbf{v},1}^{(n)} \mathbf{h}_2^{(n)} \quad \text{subject to} \quad \left(\mathbf{h}_2^{(n)}\right)^H \mathbf{d}_{2,\theta_\mathrm{d}} = 1, \tag{3.123}$$

where $\mathbf{\Phi}_{\mathbf{v},2}^{(n-1)}$ and $\mathbf{\Phi}_{\mathbf{v},1}^{(n)}$ are defined in (3.114) and (3.116), respectively. From the optimization of (3.122) and (3.123), we get

$$\mathbf{h}_1^{(n)} = \frac{\left(\mathbf{\Phi}_{\mathbf{v},2}^{(n-1)}\right)^{-1} \mathbf{d}_{1,\theta_\mathrm{d}}}{\mathbf{d}_{1,\theta_\mathrm{d}}^H \left(\mathbf{\Phi}_{\mathbf{v},2}^{(n-1)}\right)^{-1} \mathbf{d}_{1,\theta_\mathrm{d}}} \tag{3.124}$$

and

$$\mathbf{h}_2^{(n)} = \frac{\left(\mathbf{\Phi}_{\mathbf{v},1}^{(n)}\right)^{-1} \mathbf{d}_{2,\theta_\mathrm{d}}}{\mathbf{d}_{2,\theta_\mathrm{d}}^H \left(\mathbf{\Phi}_{\mathbf{v},1}^{(n)}\right)^{-1} \mathbf{d}_{2,\theta_\mathrm{d}}}, \tag{3.125}$$

with the initialization:

$$\mathbf{h}_2^{(0)} = \frac{\mathbf{\Phi}_{\mathbf{v}_2}^{-1} \mathbf{d}_{2,\theta_\mathrm{d}}}{\mathbf{d}_{2,\theta_\mathrm{d}}^H \mathbf{\Phi}_{\mathbf{v}_2}^{-1} \mathbf{d}_{2,\theta_\mathrm{d}}}. \tag{3.126}$$

As a result, the MVDR beamformer is at iteration n:

$$\mathbf{h}_{\mathrm{MVDR1}}^{(n)} = \mathbf{h}_1^{(n)} \otimes \mathbf{h}_2^{(n)}. \tag{3.127}$$

This beamformer can be directly obtained from $\mathbf{h}_{\mathrm{T},\mu_1,\mu_2}^{(n)}$ by taking $\mu_1 = \mu_2 = 0$.

Another form of the MVDR beamformer is obtained from

$$\min_{\mathbf{h}_1^{(n)}} \left(\mathbf{h}_1^{(n)}\right)^H \mathbf{\Phi}_{\mathbf{y},2}^{(n-1)} \mathbf{h}_1^{(n)} \quad \text{subject to} \quad \left(\mathbf{h}_1^{(n)}\right)^H \mathbf{d}_{1,\theta_\mathrm{d}} = 1, \tag{3.128}$$

$$\min_{\mathbf{h}_2^{(n)}} \left(\mathbf{h}_2^{(n)}\right)^H \mathbf{\Phi}_{\mathbf{y},1}^{(n)} \mathbf{h}_2^{(n)} \quad \text{subject to} \quad \left(\mathbf{h}_2^{(n)}\right)^H \mathbf{d}_{2,\theta_\mathrm{d}} = 1, \tag{3.129}$$

where $\mathbf{\Phi}_{\mathbf{y},2}^{(n-1)}$ and $\mathbf{\Phi}_{\mathbf{y},1}^{(n)}$ are defined in (3.102) and (3.105), respectively. We get

$$\mathbf{h}_{\mathrm{MVDR2}}^{(n)} = \mathbf{h}_1^{(n)} \otimes \mathbf{h}_2^{(n)}, \tag{3.130}$$

where

$$\mathbf{h}_1^{(n)} = \frac{\left(\mathbf{\Phi}_{\mathbf{y},2}^{(n-1)}\right)^{-1} \mathbf{d}_{1,\theta_\mathrm{d}}}{\mathbf{d}_{1,\theta_\mathrm{d}}^H \left(\mathbf{\Phi}_{\mathbf{y},2}^{(n-1)}\right)^{-1} \mathbf{d}_{1,\theta_\mathrm{d}}} \tag{3.131}$$

and

$$\mathbf{h}_2^{(n)} = \frac{\left(\mathbf{\Phi}_{\mathbf{y},1}^{(n)}\right)^{-1} \mathbf{d}_{2,\theta_\mathrm{d}}}{\mathbf{d}_{2,\theta_\mathrm{d}}^H \left(\mathbf{\Phi}_{\mathbf{y},1}^{(n)}\right)^{-1} \mathbf{d}_{2,\theta_\mathrm{d}}}, \tag{3.132}$$

with the initialization:

$$\mathbf{h}_2^{(0)} = \frac{\mathbf{\Phi}_{\mathbf{y}_2}^{-1} \mathbf{d}_{2,\theta_\mathrm{d}}}{\mathbf{d}_{2,\theta_\mathrm{d}}^H \mathbf{\Phi}_{\mathbf{y}_2}^{-1} \mathbf{d}_{2,\theta_\mathrm{d}}}. \tag{3.133}$$

In principle, $\mathbf{h}_{\mathrm{MVDR1}}^{(n)}$ and $\mathbf{h}_{\mathrm{MVDR2}}^{(n)}$ are equivalent, but in practice, they may behave very differently.

3.2.5 LCMV

We assume that we have one interference source impinging on the array from the direction $\theta_0 \neq \theta_\mathrm{d}$ that we would like to cancel without distorting the desired signal. Then, our constraint equations are identical to the ones presented in (3.43) and (3.44). Depending on what we wish; one null of multiplicity 1 or one null of multiplicity 2 in the direction θ_0, we can derive different linearly constrained minimum variance (LCMV) beamformers [7], [8].

If one null of multiplicity 2 is desired, we should optimize

$$\min_{\mathbf{h}_1^{(n)}} \left(\mathbf{h}_1^{(n)}\right)^H \mathbf{\Phi}_{\mathbf{v},2}^{(n-1)} \mathbf{h}_1^{(n)} \quad \text{subject to} \quad \mathbf{C}_1^H \mathbf{h}_1^{(n)} = \mathbf{i}_\mathrm{c}, \tag{3.134}$$

$$\min_{\mathbf{h}_2^{(n)}} \left(\mathbf{h}_2^{(n)}\right)^H \mathbf{\Phi}_{\mathbf{v},1}^{(n)} \mathbf{h}_2^{(n)} \quad \text{subject to} \quad \mathbf{C}_2^H \mathbf{h}_2^{(n)} = \mathbf{i}_\mathrm{c}. \tag{3.135}$$

We find that

$$\mathbf{h}_1^{(n)} = \left(\mathbf{\Phi}_{\mathbf{v},2}^{(n-1)}\right)^{-1} \mathbf{C}_1 \left[\mathbf{C}_1^H \left(\mathbf{\Phi}_{\mathbf{v},2}^{(n-1)}\right)^{-1} \mathbf{C}_1\right]^{-1} \mathbf{i}_\mathrm{c} \tag{3.136}$$

and

$$\mathbf{h}_2^{(n)} = \left(\mathbf{\Phi}_{\mathbf{v},1}^{(n)}\right)^{-1} \mathbf{C}_2 \left[\mathbf{C}_2^H \left(\mathbf{\Phi}_{\mathbf{v},1}^{(n)}\right)^{-1} \mathbf{C}_2\right]^{-1} \mathbf{i}_\mathrm{c}, \tag{3.137}$$

with the initialization:

$$\mathbf{h}_2^{(0)} = \mathbf{\Phi}_{\mathbf{v}_2}^{-1}\mathbf{C}_2 \left(\mathbf{C}_2^H \mathbf{\Phi}_{\mathbf{v}_2}^{-1}\mathbf{C}_2\right)^{-1} \mathbf{i}_{\mathrm{c}}. \quad (3.138)$$

As a consequence, the first LCMV beamformer is at iteration n:

$$\mathbf{h}_{\mathrm{LCMV1}}^{(n)} = \mathbf{h}_1^{(n)} \otimes \mathbf{h}_2^{(n)}. \quad (3.139)$$

A second LCMV beamformer (with one null of multiplicity 2) is at iteration n:

$$\mathbf{h}_{\mathrm{LCMV2}}^{(n)} = \mathbf{h}_1^{(n)} \otimes \mathbf{h}_2^{(n)}, \quad (3.140)$$

where

$$\mathbf{h}_1^{(n)} = \left(\mathbf{\Phi}_{\mathbf{y},2}^{(n-1)}\right)^{-1} \mathbf{C}_1 \left[\mathbf{C}_1^H \left(\mathbf{\Phi}_{\mathbf{y},2}^{(n-1)}\right)^{-1} \mathbf{C}_1\right]^{-1} \mathbf{i}_{\mathrm{c}} \quad (3.141)$$

and

$$\mathbf{h}_2^{(n)} = \left(\mathbf{\Phi}_{\mathbf{y},1}^{(n)}\right)^{-1} \mathbf{C}_2 \left[\mathbf{C}_2^H \left(\mathbf{\Phi}_{\mathbf{y},1}^{(n)}\right)^{-1} \mathbf{C}_2\right]^{-1} \mathbf{i}_{\mathrm{c}}, \quad (3.142)$$

with the initialization:

$$\mathbf{h}_2^{(0)} = \mathbf{\Phi}_{\mathbf{y}_2}^{-1}\mathbf{C}_2 \left(\mathbf{C}_2^H \mathbf{\Phi}_{\mathbf{y}_2}^{-1}\mathbf{C}_2\right)^{-1} \mathbf{i}_{\mathrm{c}}. \quad (3.143)$$

If one null of multiplicity 1 is desired, we can optimize

$$\min_{\mathbf{h}_1^{(n)}} \left(\mathbf{h}_1^{(n)}\right)^H \mathbf{\Phi}_{\mathbf{v},2}^{(n-1)} \mathbf{h}_1^{(n)} \quad \text{subject to} \quad \mathbf{d}_{1,\theta_{\mathrm{d}}}^H \mathbf{h}_1^{(n)} = 1, \quad (3.144)$$

$$\min_{\mathbf{h}_2^{(n)}} \left(\mathbf{h}_2^{(n)}\right)^H \mathbf{\Phi}_{\mathbf{v},1}^{(n)} \mathbf{h}_2^{(n)} \quad \text{subject to} \quad \mathbf{C}_2^H \mathbf{h}_2^{(n)} = \mathbf{i}_{\mathrm{c}}. \quad (3.145)$$

We find that

$$\mathbf{h}_1^{(n)} = \frac{\left(\mathbf{\Phi}_{\mathbf{v},2}^{(n-1)}\right)^{-1} \mathbf{d}_{1,\theta_{\mathrm{d}}}}{\mathbf{d}_{1,\theta_{\mathrm{d}}}^H \left(\mathbf{\Phi}_{\mathbf{v},2}^{(n-1)}\right)^{-1} \mathbf{d}_{1,\theta_{\mathrm{d}}}} \quad (3.146)$$

and

$$\mathbf{h}_2^{(n)} = \left(\mathbf{\Phi}_{\mathbf{v},1}^{(n)}\right)^{-1} \mathbf{C}_2 \left[\mathbf{C}_2^H \left(\mathbf{\Phi}_{\mathbf{v},1}^{(n)}\right)^{-1} \mathbf{C}_2\right]^{-1} \mathbf{i}_{\mathrm{c}}, \quad (3.147)$$

with the initialization:

$$\mathbf{h}_2^{(0)} = \mathbf{\Phi}_{\mathbf{v}_2}^{-1}\mathbf{C}_2\left(\mathbf{C}_2^H\mathbf{\Phi}_{\mathbf{v}_2}^{-1}\mathbf{C}_2\right)^{-1}\mathbf{i}_{\mathrm{c}}. \tag{3.148}$$

Therefore, the third LCMV beamformer, but with one null of multiplicity 1, is at iteration n:

$$\mathbf{h}_{\mathrm{LCMV3}}^{(n)} = \mathbf{h}_1^{(n)} \otimes \mathbf{h}_2^{(n)}. \tag{3.149}$$

The fourth LCMV beamformer with one null of multiplicity 1 is at iteration n:

$$\mathbf{h}_{\mathrm{LCMV4}}^{(n)} = \mathbf{h}_1^{(n)} \otimes \mathbf{h}_2^{(n)}, \tag{3.150}$$

where

$$\mathbf{h}_1^{(n)} = \frac{\left(\mathbf{\Phi}_{\mathbf{y},2}^{(n-1)}\right)^{-1}\mathbf{d}_{1,\theta_{\mathrm{d}}}}{\mathbf{d}_{1,\theta_{\mathrm{d}}}^H\left(\mathbf{\Phi}_{\mathbf{y},2}^{(n-1)}\right)^{-1}\mathbf{d}_{1,\theta_{\mathrm{d}}}} \tag{3.151}$$

and

$$\mathbf{h}_2^{(n)} = \left(\mathbf{\Phi}_{\mathbf{y},1}^{(n)}\right)^{-1}\mathbf{C}_2\left[\mathbf{C}_2^H\left(\mathbf{\Phi}_{\mathbf{y},1}^{(n)}\right)^{-1}\mathbf{C}_2\right]^{-1}\mathbf{i}_{\mathrm{c}}, \tag{3.152}$$

with the initialization:

$$\mathbf{h}_2^{(0)} = \mathbf{\Phi}_{\mathbf{y}_2}^{-1}\mathbf{C}_2\left(\mathbf{C}_2^H\mathbf{\Phi}_{\mathbf{y}_2}^{-1}\mathbf{C}_2\right)^{-1}\mathbf{i}_{\mathrm{c}}. \tag{3.153}$$

Another possibility is to optimize

$$\min_{\mathbf{h}_1^{(n)}}\left(\mathbf{h}_1^{(n)}\right)^H\mathbf{\Phi}_{\mathbf{v},2}^{(n-1)}\mathbf{h}_1^{(n)} \quad \text{subject to} \quad \mathbf{C}_1^H\mathbf{h}_1^{(n)} = \mathbf{i}_{\mathrm{c}}, \tag{3.154}$$

$$\min_{\mathbf{h}_2^{(n)}}\left(\mathbf{h}_2^{(n)}\right)^H\mathbf{\Phi}_{\mathbf{v},1}^{(n)}\mathbf{h}_2^{(n)} \quad \text{subject to} \quad \mathbf{d}_{2,\theta_{\mathrm{d}}}^H\mathbf{h}_2^{(n)} = 1. \tag{3.155}$$

In this case, we get

$$\mathbf{h}_1^{(n)} = \left(\mathbf{\Phi}_{\mathbf{v},2}^{(n-1)}\right)^{-1}\mathbf{C}_1\left[\mathbf{C}_1^H\left(\mathbf{\Phi}_{\mathbf{v},2}^{(n-1)}\right)^{-1}\mathbf{C}_1\right]^{-1}\mathbf{i}_{\mathrm{c}} \tag{3.156}$$

and

$$\mathbf{h}_2^{(n)} = \frac{\left(\mathbf{\Phi}_{\mathbf{v},1}^{(n)}\right)^{-1}\mathbf{d}_{2,\theta_{\mathrm{d}}}}{\mathbf{d}_{2,\theta_{\mathrm{d}}}^H\left(\mathbf{\Phi}_{\mathbf{v},1}^{(n)}\right)^{-1}\mathbf{d}_{2,\theta_{\mathrm{d}}}}, \tag{3.157}$$

with the initialization:

$$\mathbf{h}_2^{(0)} = \frac{\mathbf{\Phi}_{\mathbf{v}_2}^{-1}\mathbf{d}_{2,\theta_\mathrm{d}}}{\mathbf{d}_{2,\theta_\mathrm{d}}^H \mathbf{\Phi}_{\mathbf{v}_2}^{-1}\mathbf{d}_{2,\theta_\mathrm{d}}}. \tag{3.158}$$

Therefore, the fifth LCMV beamformer, but with one null of multiplicity 1, is at iteration n:

$$\mathbf{h}_{\mathrm{LCMV5}}^{(n)} = \mathbf{h}_1^{(n)} \otimes \mathbf{h}_2^{(n)}. \tag{3.159}$$

Finally, the last and sixth LCMV beamformer (with one null of multiplicity 1) is at iteration n:

$$\mathbf{h}_{\mathrm{LCMV6}}^{(n)} = \mathbf{h}_1^{(n)} \otimes \mathbf{h}_2^{(n)}, \tag{3.160}$$

where

$$\mathbf{h}_1^{(n)} = \left(\mathbf{\Phi}_{\mathbf{y},2}^{(n-1)}\right)^{-1} \mathbf{C}_1 \left[\mathbf{C}_1^H \left(\mathbf{\Phi}_{\mathbf{y},2}^{(n-1)}\right)^{-1} \mathbf{C}_1\right]^{-1} \mathbf{i}_\mathrm{c} \tag{3.161}$$

and

$$\mathbf{h}_2^{(n)} = \frac{\left(\mathbf{\Phi}_{\mathbf{y},1}^{(n)}\right)^{-1} \mathbf{d}_{2,\theta_\mathrm{d}}}{\mathbf{d}_{2,\theta_\mathrm{d}}^H \left(\mathbf{\Phi}_{\mathbf{y},1}^{(n)}\right)^{-1} \mathbf{d}_{2,\theta_\mathrm{d}}}, \tag{3.162}$$

with the initialization:

$$\mathbf{h}_2^{(0)} = \frac{\mathbf{\Phi}_{\mathbf{y}_2}^{-1}\mathbf{d}_{2,\theta_\mathrm{d}}}{\mathbf{d}_{2,\theta_\mathrm{d}}^H \mathbf{\Phi}_{\mathbf{y}_2}^{-1}\mathbf{d}_{2,\theta_\mathrm{d}}}. \tag{3.163}$$

3.2.6 Maximum SNR

In order to maximize the output SNR, it is required to express it differently. Indeed, from its definition, it is clear that it can be rewritten as

$$\begin{aligned}
\mathrm{oSNR}\left(\mathbf{h}_1 \otimes \mathbf{h}_2\right) &= \frac{\phi_X \left|\mathbf{h}_1^H \mathbf{d}_{1,\theta_\mathrm{d}}\right|^2 \left|\mathbf{h}_2^H \mathbf{d}_{2,\theta_\mathrm{d}}\right|^2}{\left(\mathbf{h}_1 \otimes \mathbf{h}_2\right)^H \mathbf{\Phi}_\mathbf{v} \left(\mathbf{h}_1 \otimes \mathbf{h}_2\right)} \\
&= \frac{\left|\phi_{X,2}\right|^2 \left|\mathbf{h}_1^H \mathbf{d}_{1,\theta_\mathrm{d}}\right|^2}{\phi_X \mathbf{h}_1^H \mathbf{\Phi}_{\mathbf{v},2} \mathbf{h}_1} \qquad (3.164) \\
&= \frac{\left|\phi_{X,1}\right|^2 \left|\mathbf{h}_2^H \mathbf{d}_{2,\theta_\mathrm{d}}\right|^2}{\phi_X \mathbf{h}_2^H \mathbf{\Phi}_{\mathbf{v},1} \mathbf{h}_2}, \qquad (3.165)
\end{aligned}$$

where

$$|\phi_{X,2}|^2 = \phi_X^2 \left|\mathbf{h}_2^H \mathbf{d}_{2,\theta_d}\right|^2, \tag{3.166}$$

$$\mathbf{\Phi}_{\mathbf{v},2} = (\mathbf{I}_{M_0} \otimes \mathbf{h}_2)^H \mathbf{\Phi}_{\mathbf{v}} (\mathbf{I}_{M_0} \otimes \mathbf{h}_2), \tag{3.167}$$

and

$$|\phi_{X,1}|^2 = \phi_X^2 \left|\mathbf{h}_1^H \mathbf{d}_{1,\theta_d}\right|^2, \tag{3.168}$$

$$\mathbf{\Phi}_{\mathbf{v},1} = (\mathbf{h}_1 \otimes \mathbf{I}_{M_0})^H \mathbf{\Phi}_{\mathbf{v}} (\mathbf{h}_1 \otimes \mathbf{I}_{M_0}). \tag{3.169}$$

When $\mathbf{h}_2$ is fixed, we write (3.164) as

$$\text{oSNR}(\mathbf{h}_1|\mathbf{h}_2) = \frac{|\phi_{X,2}|^2 \left|\mathbf{h}_1^H \mathbf{d}_{1,\theta_d}\right|^2}{\phi_X \mathbf{h}_1^H \mathbf{\Phi}_{\mathbf{v},2} \mathbf{h}_1}, \tag{3.170}$$

and when $\mathbf{h}_1$ is fixed, we write (3.165) as

$$\text{oSNR}(\mathbf{h}_2|\mathbf{h}_1) = \frac{|\phi_{X,1}|^2 \left|\mathbf{h}_2^H \mathbf{d}_{2,\theta_d}\right|^2}{\phi_X \mathbf{h}_2^H \mathbf{\Phi}_{\mathbf{v},1} \mathbf{h}_2}. \tag{3.171}$$

As before, we have everything to iteratively maximize the output SNR. At iteration 0, we may take

$$\mathbf{h}_2^{(0)} = \alpha_2^{(0)} \mathbf{\Phi}_{\mathbf{v}_2}^{-1} \mathbf{d}_{2,\theta_d}, \tag{3.172}$$

where $\alpha_2^{(0)} \neq 0$ is an arbitrary complex-valued number. Substituting $\mathbf{h}_2^{(0)}$ into (3.166)–(3.167), we obtain

$$\left|\phi_{X,2}^{(0)}\right|^2 = \phi_X^2 \left|\mathbf{d}_{2,\theta_d}^H \mathbf{h}_2^{(0)}\right|^2, \tag{3.173}$$

$$\mathbf{\Phi}_{\mathbf{v},2}^{(0)} = \left(\mathbf{I}_{M_0} \otimes \mathbf{h}_2^{(0)}\right)^H \mathbf{\Phi}_{\mathbf{v}} \left(\mathbf{I}_{M_0} \otimes \mathbf{h}_2^{(0)}\right). \tag{3.174}$$

Using the previous expressions in the output SNR in (3.170), we get at iteration 1:

$$\text{oSNR}\left(\mathbf{h}_1^{(1)}|\mathbf{h}_2^{(0)}\right) = \frac{\left|\phi_{X,2}^{(0)}\right|^2 \left|\mathbf{d}_{1,\theta_d}^H \mathbf{h}_1^{(1)}\right|^2}{\phi_X \left(\mathbf{h}_1^{(1)}\right)^H \mathbf{\Phi}_{\mathbf{v},2}^{(0)} \mathbf{h}_1^{(1)}}, \tag{3.175}$$

whose maximization with respect to $\mathbf{h}_1^{(1)}$ gives

$$\mathbf{h}_1^{(1)} = \alpha_1^{(0)} \left(\mathbf{\Phi}_{\mathbf{v},2}^{(0)}\right)^{-1} \mathbf{d}_{1,\theta_d}, \tag{3.176}$$

where $\alpha_1^{(0)} \neq 0$ is an arbitrary complex number. Then, using $\mathbf{h}_1^{(1)}$ in (3.168)–(3.169), we obtain

$$\left|\phi_{X,1}^{(1)}\right|^2 = \phi_X^2 \left|\mathbf{d}_{1,\theta_\mathrm{d}}^H \mathbf{h}_1^{(1)}\right|^2, \tag{3.177}$$

$$\mathbf{\Phi}_{\mathbf{v},1}^{(1)} = \left(\mathbf{h}_1^{(1)} \otimes \mathbf{I}_{M_0}\right)^H \mathbf{\Phi}_\mathbf{v} \left(\mathbf{h}_1^{(1)} \otimes \mathbf{I}_{M_0}\right), \tag{3.178}$$

which we plug into the output SNR in (3.171). We get

$$\mathrm{oSNR}\left(\mathbf{h}_2^{(1)}|\mathbf{h}_1^{(1)}\right) = \frac{\left|\phi_{X,1}^{(1)}\right|^2 \left|\mathbf{d}_{2,\theta_\mathrm{d}}^H \mathbf{h}_2^{(1)}\right|^2}{\phi_X \left(\mathbf{h}_2^{(1)}\right)^H \mathbf{\Phi}_{\mathbf{v},1}^{(1)} \mathbf{h}_2^{(1)}}. \tag{3.179}$$

The maximization of the previous expression with respect to $\mathbf{h}_2^{(1)}$ leads to

$$\mathbf{h}_2^{(1)} = \alpha_2^{(1)} \left(\mathbf{\Phi}_{\mathbf{v},1}^{(1)}\right)^{-1} \mathbf{d}_{2,\theta_\mathrm{d}}, \tag{3.180}$$

where $\alpha_2^{(1)} \neq 0$ is an arbitrary complex number.

Continuing the iterations up to the iteration n, we easily get for the first filter:

$$\mathbf{h}_1^{(n)} = \alpha_1^{(n-1)} \left(\mathbf{\Phi}_{\mathbf{v},2}^{(n-1)}\right)^{-1} \mathbf{d}_{1,\theta_\mathrm{d}}, \tag{3.181}$$

where $\alpha_1^{(n-1)} \neq 0$ is an arbitrary complex number, with

$$\mathbf{\Phi}_{\mathbf{v},2}^{(n-1)} = \left(\mathbf{I}_{M_0} \otimes \mathbf{h}_2^{(n-1)}\right)^H \mathbf{\Phi}_\mathbf{v} \left(\mathbf{I}_{M_0} \otimes \mathbf{h}_2^{(n-1)}\right), \tag{3.182}$$

and for the second filter:

$$\mathbf{h}_2^{(n)} = \alpha_2^{(n)} \left(\mathbf{\Phi}_{\mathbf{v},1}^{(n)}\right)^{-1} \mathbf{d}_{2,\theta_\mathrm{d}}, \tag{3.183}$$

where $\alpha_2^{(n)} \neq 0$ is an arbitrary complex number, with

$$\mathbf{\Phi}_{\mathbf{v},1}^{(n)} = \left(\mathbf{h}_1^{(n)} \otimes \mathbf{I}_{M_0}\right)^H \mathbf{\Phi}_\mathbf{v} \left(\mathbf{h}_1^{(n)} \otimes \mathbf{I}_{M_0}\right). \tag{3.184}$$

Finally, we deduce that the maximum SNR beamformer is at iteration n:

$$\mathbf{h}_{\max}^{(n)} = \mathbf{h}_1^{(n)} \otimes \mathbf{h}_2^{(n)}, \tag{3.185}$$

where $\mathbf{h}_1^{(n)}$ and $\mathbf{h}_2^{(n)}$ are defined in (3.181) and (3.183), respectively. There are different ways to derive the parameters $\alpha_1^{(n-1)}$ and $\alpha_2^{(n)}$. For example, if they are found in such a way that the filters $\mathbf{h}_1^{(n)}$ and $\mathbf{h}_2^{(n)}$ are distortionless, we obtain the MVDR beamformer derived in Subsection 3.2.4.

3.3 Combined Fixed/Adaptive Beamformers

Perhaps, one of the most interesting aspects of the Kronecker product formulation in beamforming (i.e., $\mathbf{h} = \mathbf{h}_1 \otimes \mathbf{h}_2$) is that, due to its decomposition and remarkable flexibility, it seems possible to combine very intelligently fixed and adaptive beamformers, so that the best of each one of these two approaches is emphasized for performance enhancement. Therefore, we hope that the global beamformer will fix the shortcomings of each one of these two classes of beamformers.

For the second ULA with the steering vector $\mathbf{d}_{2,\theta_\mathrm{d}}$, the interelement spacing, δ, can be chosen as small as desired, which is obviously good for directivity (especially at the endfires) and also good for limiting the effect of spatial aliasing. Therefore, the corresponding beamformer, $\mathbf{h}_2$, will be the fixed beamformer here and, in all this part, it will be taken as

$$\begin{aligned} \mathbf{h}_2 &= \mathbf{h}_{2,\mathrm{mDF}} \\ &= \frac{\mathbf{\Gamma}_2^{-1}\mathbf{d}_{2,\theta_\mathrm{d}}}{\mathbf{d}_{2,\theta_\mathrm{d}}^H \mathbf{\Gamma}_2^{-1}\mathbf{d}_{2,\theta_\mathrm{d}}}, \end{aligned} \tag{3.186}$$

which is the maximum DF beamformer at the second ULA. Now, that $\mathbf{h}_2$ is fixed, we need to derive, accordingly, the optimal (depending on what is desired) adaptive beamformer, $\mathbf{h}_1$, which corresponds to the first ULA. Clearly, the global beamformer, $\mathbf{h}$, will inherit the features of the adaptive and fixed beamformers ($\mathbf{h}_1$ and $\mathbf{h}_2$, respectively).

Since $\mathbf{h}_2$ is fixed and distortionless, we can express the MSE as

$$J(\mathbf{h}_1|\mathbf{h}_2) = \phi_X + \mathbf{h}_1^H \mathbf{\Phi}_{\mathbf{y},2}\mathbf{h}_1 - \phi_X \mathbf{h}_1^H \mathbf{d}_{1,\theta_\mathrm{d}} - \phi_X \mathbf{d}_{1,\theta_\mathrm{d}}^H \mathbf{h}_1, \tag{3.187}$$

where

$$\mathbf{\Phi}_{\mathbf{y},2} = \left(\mathbf{I}_{M_0} \otimes \mathbf{h}_{2,\mathrm{mDF}}\right)^H \mathbf{\Phi}_{\mathbf{y}} \left(\mathbf{I}_{M_0} \otimes \mathbf{h}_{2,\mathrm{mDF}}\right). \tag{3.188}$$

The minimization of $J(\mathbf{h}_1|\mathbf{h}_2)$ with respect to $\mathbf{h}_1$ leads to the Wiener (adaptive) beamformer:

$$\begin{aligned} \mathbf{h}_{1,\mathrm{W}} &= \phi_X \mathbf{\Phi}_{\mathbf{y},2}^{-1}\mathbf{d}_{1,\theta_\mathrm{d}} \\ &= \frac{\phi_X \mathbf{\Phi}_{\mathbf{v},2}^{-1}\mathbf{d}_{1,\theta_\mathrm{d}}}{1 + \phi_X \mathbf{d}_{1,\theta_\mathrm{d}}^H \mathbf{\Phi}_{\mathbf{v},2}^{-1}\mathbf{d}_{1,\theta_\mathrm{d}}}, \end{aligned} \tag{3.189}$$

where

$$\mathbf{\Phi}_{\mathbf{v},2} = \left(\mathbf{I}_{M_0} \otimes \mathbf{h}_{2,\mathrm{mDF}}\right)^H \mathbf{\Phi}_{\mathbf{v}} \left(\mathbf{I}_{M_0} \otimes \mathbf{h}_{2,\mathrm{mDF}}\right). \tag{3.190}$$

As a result, the first proposed global combined fixed/adaptive (cFA) beamformer is

$$\mathbf{h}_{\mathrm{cFA1}} = \mathbf{h}_{1,\mathrm{W}} \otimes \mathbf{h}_{2,\mathrm{mDF}}. \tag{3.191}$$

It is always possible to improve the beamforming performance by adding one more step to the previous processing, i.e, by maximizing

$$\mathcal{D}\left(\mathbf{h}_2 | \mathbf{h}_{1,\mathrm{W}}\right) = \frac{\left|\mathbf{h}_2^H \mathbf{d}_{2,\theta_\mathrm{d}}\right|^2}{\mathbf{h}_2^H \mathbf{\Gamma}_{\mathbf{h}_{1,\mathrm{W}}} \mathbf{h}_2}, \tag{3.192}$$

where

$$\mathbf{\Gamma}_{\mathbf{h}_{1,\mathrm{W}}} = \left(\mathbf{h}_{1,\mathrm{W}} \otimes \mathbf{I}_{M_0}\right)^H \mathbf{\Gamma} \left(\mathbf{h}_{1,\mathrm{W}} \otimes \mathbf{I}_{M_0}\right).$$

We obtain

$$\mathbf{h}_{2,\mathrm{mDF}}^{(1)} = \frac{\mathbf{\Gamma}_{\mathbf{h}_{1,\mathrm{W}}}^{-1} \mathbf{d}_{2,\theta_\mathrm{d}}}{\mathbf{d}_{2,\theta_\mathrm{d}}^H \mathbf{\Gamma}_{\mathbf{h}_{1,\mathrm{W}}}^{-1} \mathbf{d}_{2,\theta_\mathrm{d}}}. \tag{3.193}$$

Then, instead of using $\mathbf{h}_{2,\mathrm{mDF}}$ in (3.191), we can use $\mathbf{h}_{2,\mathrm{mDF}}^{(1)}$. This leads to the second global beamformer:

$$\mathbf{h}_{\mathrm{cFA2}} = \mathbf{h}_{1,\mathrm{W}} \otimes \mathbf{h}_{2,\mathrm{mDF}}^{(1)}. \tag{3.194}$$

If we want $\mathbf{h}_1$ to be a distortionless adaptive beamformer, we can optimize the criterion:

$$\min_{\mathbf{h}_1} \mathbf{h}_1^H \mathbf{\Phi}_{\mathbf{v},2} \mathbf{h}_1 \quad \text{subject to} \quad \mathbf{h}_1^H \mathbf{d}_{1,\theta_\mathrm{d}} = 1, \tag{3.195}$$

from which we find the well-known MVDR beamformer:

$$\mathbf{h}_{1,\mathrm{MVDR}} = \frac{\mathbf{\Phi}_{\mathbf{v},2}^{-1} \mathbf{d}_{1,\theta_\mathrm{d}}}{\mathbf{d}_{1,\theta_\mathrm{d}}^H \mathbf{\Phi}_{\mathbf{v},2}^{-1} \mathbf{d}_{1,\theta_\mathrm{d}}}. \tag{3.196}$$

As a consequence, the third proposed global fixed/adaptive beamformer is

$$\mathbf{h}_{\mathrm{cFA3}} = \mathbf{h}_{1,\mathrm{MVDR}} \otimes \mathbf{h}_{2,\mathrm{mDF}}. \tag{3.197}$$

In the particular case where the noise is white, $\mathbf{h}_{1,\mathrm{MVDR}}$ simplifies to the DS beamformer, i.e., $\mathbf{h}_{1,\mathrm{DS}}$. Therefore, $\mathbf{h}_{\mathrm{cFA3}} = \mathbf{h}_{1,\mathrm{DS}} \otimes \mathbf{h}_{2,\mathrm{mDF}} = \mathbf{h}_{\mathrm{PmDF6}}$ (see Subsection 3.1.2).

If we want to better compromise between noise reduction and speech distortion, we should optimize the criterion:

$$\min_{\mathbf{h}_1} J_\mathrm{d}\left(\mathbf{h}_1 | \mathbf{h}_2\right) \quad \text{subject to} \quad J_\mathrm{n}\left(\mathbf{h}_1 | \mathbf{h}_2\right) = \aleph_1 \phi_{V_1}, \tag{3.198}$$

where $0 < \aleph_1 < 1$ and

$$J_{\rm d}\left(\mathbf{h}_1|\mathbf{h}_2\right) = \phi_X + \phi_X \left|\mathbf{h}_1^H \mathbf{d}_{1,\theta_{\rm d}}\right|^2 - \phi_X \mathbf{h}_1^H \mathbf{d}_{1,\theta_{\rm d}} - \phi_X \mathbf{d}_{1,\theta_{\rm d}}^H \mathbf{h}_1, \quad (3.199)$$

$$J_{\rm n}\left(\mathbf{h}_1|\mathbf{h}_2\right) = \mathbf{h}_1^H \mathbf{\Phi}_{\mathbf{v},2} \mathbf{h}_1. \quad (3.200)$$

We get the tradeoff beamformer:

$$\begin{aligned} \mathbf{h}_{1,\mathrm{T},\mu_1} &= \phi_X \left(\phi_X \mathbf{d}_{1,\theta_{\rm d}} \mathbf{d}_{1,\theta_{\rm d}}^H + \mu_1 \mathbf{\Phi}_{\mathbf{v},2}\right)^{-1} \mathbf{d}_{1,\theta_{\rm d}} \\ &= \frac{\phi_X \mathbf{\Phi}_{\mathbf{v},2}^{-1} \mathbf{d}_{1,\theta_{\rm d}}}{\mu_1 + \phi_X \mathbf{d}_{1,\theta_{\rm d}}^H \mathbf{\Phi}_{\mathbf{v},2}^{-1} \mathbf{d}_{1,\theta_{\rm d}}}, \end{aligned} \quad (3.201)$$

where $\mu_1 > 0$ is a Lagrange multiplier. Therefore, the fourth and last global fixed/adaptive beamformer that we propose is

$$\mathbf{h}_{\rm cFA4} = \mathbf{h}_{1,\mathrm{T},\mu_1} \otimes \mathbf{h}_{2,\mathrm{mDF}}. \quad (3.202)$$

3.4 Differential Beamformers

The family of differential beamformers is an important particular class of fixed beamformers. Arguably, beamformers belonging to this particular family are the most practical ones since the corresponding beampatterns are almost frequency invariant, which is critical when we deal with broadband signals such as speech, and they lead to the highest gains in diffuse noise. However, the main drawback of differential beamforming is white noise amplification. We will see that the flexibility of the new approach allows us to better deal with this fundamental problem. The most well-known and studied differential array beampatterns are the cardioid, the dipole, the hypercardioid, and the supercardioid. In the following, we show how they are designed in this particular context.

3.4.1 Preliminaries and Other Measures

In order that differential beamforming takes place, the following two assumptions are usually made [9], [10], [11], [12].

(i) The sensor spacing, δ, is much smaller than the acoustic wavelength, implying that $\delta \ll 2\pi c/\omega$. This assumption is required so that the true acoustic pressure differentials can be approximated by finite differences of the sensors' outputs.

(ii) The desired source signal propagates from the angle $\theta_{\rm d} = 0$ (endfire direction). Therefore, (2.6) becomes

$$\mathbf{y} = \mathbf{d}_0 X + \mathbf{v}, \quad (3.203)$$

and, at the endfire, the value of the beamformer beampattern should always be equal to 1.

Thanks to Assumption (i) frequency-invariant beamforming may be possible and thanks to Assumption (ii) any desired beampattern can be designed; in other directions (than the endfires 0 and π), the beampattern design is very limited because of the symmetry of the steering vector.

Since the interelement spacing of the second ULA is much smaller than the interelement spacing of the first ULA, the filter $\mathbf{h}_2$ will be used to design and shape the directivity pattern while $\mathbf{h}_1$ will be used to mostly maximize the WNG. Consequently, with $\mathbf{h}_2$, we will design an $(M_0 - 1)$th-order differential beamformer. Then, the global ULA will be a differential array of order certainly higher than $M_0 - 1$.

We recall that the definitions of the WNG and the DF are, respectively,

$$\begin{aligned} \mathcal{W}(\mathbf{h}) &= \frac{\left|\mathbf{h}^H \mathbf{d}_0\right|^2}{\mathbf{h}^H \mathbf{h}} \\ &= \frac{\left|\mathbf{h}_1^H \mathbf{d}_{1,0}\right|^2}{\mathbf{h}_1^H \mathbf{h}_1} \times \frac{\left|\mathbf{h}_2^H \mathbf{d}_{2,0}\right|^2}{\mathbf{h}_2^H \mathbf{h}_2} \end{aligned} \tag{3.204}$$

and

$$\begin{aligned} \mathcal{D}(\mathbf{h}) &= \frac{\left|\mathcal{B}_0(\mathbf{h})\right|^2}{\frac{1}{2}\int_0^{\pi} \left|\mathcal{B}_\theta(\mathbf{h})\right|^2 \sin\theta d\theta} \\ &= \frac{\left|\mathcal{B}_{1,0}(\mathbf{h}_1)\right|^2 \left|\mathcal{B}_{2,0}(\mathbf{h}_2)\right|^2}{\frac{1}{2}\int_0^{\pi} \left|\mathcal{B}_{1,\theta}(\mathbf{h}_1)\right|^2 \left|\mathcal{B}_{2,\theta}(\mathbf{h}_2)\right|^2 \sin\theta d\theta} \\ &= \frac{\left|\mathbf{h}^H \mathbf{d}_0\right|^2}{\mathbf{h}^H \boldsymbol{\Gamma} \mathbf{h}}. \end{aligned} \tag{3.205}$$

Another measure of interest in this study is the front-to-back ratio (FBR), which is defined as the ratio of the power of the output of the array to signals propagating from the front-half plane to the output power for signals arriving from the rear-half plane [13]. This ratio, for the spherically isotropic (diffuse) noise field, is mathematically defined as [13]

$$\mathcal{F}(\mathbf{h}) = \frac{\int_0^{\pi/2} |\mathcal{B}_\theta(\mathbf{h})|^2 \sin\theta d\theta}{\int_{\pi/2}^{\pi} |\mathcal{B}_\theta(\mathbf{h})|^2 \sin\theta d\theta} \tag{3.206}$$
$$= \frac{\int_0^{\pi/2} |\mathcal{B}_{1,\theta}(\mathbf{h}_1)|^2 |\mathcal{B}_{2,\theta}(\mathbf{h}_2)|^2 \sin\theta d\theta}{\int_{\pi/2}^{\pi} |\mathcal{B}_{1,\theta}(\mathbf{h}_1)|^2 |\mathcal{B}_{2,\theta}(\mathbf{h}_2)|^2 \sin\theta d\theta}$$
$$= \frac{\mathbf{h}^H \mathbf{\Gamma}_\mathrm{f} \mathbf{h}}{\mathbf{h}^H \mathbf{\Gamma}_\mathrm{b} \mathbf{h}},$$

where

$$\mathbf{\Gamma}_\mathrm{f} = \int_0^{\pi/2} \mathbf{d}_\theta \mathbf{d}_\theta^H \sin\theta d\theta, \tag{3.207}$$
$$\mathbf{\Gamma}_\mathrm{b} = \int_{\pi/2}^{\pi} \mathbf{d}_\theta \mathbf{d}_\theta^H \sin\theta d\theta. \tag{3.208}$$

It can be verified that the elements of the $M \times M$ matrices $\mathbf{\Gamma}_\mathrm{f}(\omega)$ and $\mathbf{\Gamma}_\mathrm{b}(\omega)$ are given, respectively, by

$$[\mathbf{\Gamma}_\mathrm{f}(\omega)]_{ij} = \frac{e^{\jmath\omega(j-i)\delta/c} - 1}{\jmath\omega(j-i)\delta/c} \tag{3.209}$$

and

$$[\mathbf{\Gamma}_\mathrm{b}(\omega)]_{ij} = \frac{1 - e^{-\jmath\omega(j-i)\delta/c}}{\jmath\omega(j-i)\delta/c}, \tag{3.210}$$

with $[\mathbf{\Gamma}_\mathrm{f}(\omega)]_{mm} = [\mathbf{\Gamma}_\mathrm{b}(\omega)]_{mm} = 1,\ m = 1, 2, \ldots, M$. Same as the DF, the FBR cannot be factorized, i.e.,

$$\mathcal{F}(\mathbf{h}) \neq \mathcal{F}_1(\mathbf{h}_1) \times \mathcal{F}_2(\mathbf{h}_2), \tag{3.211}$$

where

$$\mathcal{F}_1(\mathbf{h}_1) = \frac{\int_0^{\pi/2} |\mathcal{B}_{1,\theta}(\mathbf{h}_1)|^2 \sin\theta d\theta}{\int_{\pi/2}^{\pi} |\mathcal{B}_{1,\theta}(\mathbf{h}_1)|^2 \sin\theta d\theta} = \frac{\mathbf{h}_1^H \mathbf{\Gamma}_{\mathrm{f},1} \mathbf{h}_1}{\mathbf{h}_1^H \mathbf{\Gamma}_{\mathrm{b},1} \mathbf{h}_1}, \tag{3.212}$$

$$\mathcal{F}_2(\mathbf{h}_2) = \frac{\int_0^{\pi/2} |\mathcal{B}_{2,\theta}(\mathbf{h}_2)|^2 \sin\theta d\theta}{\int_{\pi/2}^{\pi} |\mathcal{B}_{2,\theta}(\mathbf{h}_2)|^2 \sin\theta d\theta} = \frac{\mathbf{h}_2^H \mathbf{\Gamma}_{\mathrm{f},2} \mathbf{h}_2}{\mathbf{h}_2^H \mathbf{\Gamma}_{\mathrm{b},2} \mathbf{h}_2}, \tag{3.213}$$

with

$$\mathbf{\Gamma}_{\mathrm{f},1} = \int_0^{\pi/2} \mathbf{d}_{1,\theta} \mathbf{d}_{1,\theta}^H \sin\theta d\theta, \tag{3.214}$$

$$\mathbf{\Gamma}_{\mathrm{b},1} = \int_{\pi/2}^{\pi} \mathbf{d}_{1,\theta} \mathbf{d}_{1,\theta}^H \sin\theta d\theta, \tag{3.215}$$

$$\mathbf{\Gamma}_{\mathrm{f},2} = \int_0^{\pi/2} \mathbf{d}_{2,\theta} \mathbf{d}_{2,\theta}^H \sin\theta d\theta, \tag{3.216}$$

$$\mathbf{\Gamma}_{\mathrm{b},2} = \int_{\pi/2}^{\pi} \mathbf{d}_{2,\theta} \mathbf{d}_{2,\theta}^H \sin\theta d\theta. \tag{3.217}$$

The elements of the $M_0 \times M_0$ matrices $\mathbf{\Gamma}_{\mathrm{f},1}(\omega)$, $\mathbf{\Gamma}_{\mathrm{b},1}(\omega)$, $\mathbf{\Gamma}_{\mathrm{f},2}(\omega)$, and $\mathbf{\Gamma}_{\mathrm{b},2}(\omega)$ are given, respectively, by

$$[\mathbf{\Gamma}_{\mathrm{f},1}(\omega)]_{ij} = \frac{e^{\jmath\omega(j-i)M_0\delta/c} - 1}{\jmath\omega(j-i)M_0\delta/c}, \tag{3.218}$$

$$[\mathbf{\Gamma}_{\mathrm{b},1}(\omega)]_{ij} = \frac{1 - e^{-\jmath\omega(j-i)M_0\delta/c}}{\jmath\omega(j-i)M_0\delta/c}, \tag{3.219}$$

$$[\mathbf{\Gamma}_{\mathrm{f},2}(\omega)]_{ij} = \frac{e^{\jmath\omega(j-i)\delta/c} - 1}{\jmath\omega(j-i)\delta/c}, \tag{3.220}$$

and

$$[\mathbf{\Gamma}_{\mathrm{b},2}(\omega)]_{ij} = \frac{1 - e^{-\jmath\omega(j-i)\delta/c}}{\jmath\omega(j-i)\delta/c}, \tag{3.221}$$

with $[\mathbf{\Gamma}_{\mathrm{f},1}(\omega)]_{mm} = [\mathbf{\Gamma}_{\mathrm{b},1}(\omega)]_{mm} = [\mathbf{\Gamma}_{\mathrm{f},2}(\omega)]_{mm} = [\mathbf{\Gamma}_{\mathrm{b},2}(\omega)]_{mm} = 1$, $m = 1, 2, \ldots, M_0$.

When $\mathbf{h}_2$ is fixed and given, and thanks to (2.41), we can write the FBR as

$$\mathcal{F}(\mathbf{h}_1|\mathbf{h}_2) = \frac{\mathbf{h}_1^H \mathbf{\Gamma}_{\mathrm{f},\mathbf{h}_2} \mathbf{h}_1}{\mathbf{h}_1^H \mathbf{\Gamma}_{\mathrm{b},\mathbf{h}_2} \mathbf{h}_1}, \tag{3.222}$$

where

$$\begin{aligned}\mathbf{\Gamma}_{\mathrm{f},\mathbf{h}_2} &= \int_0^{\pi/2} \mathbf{d}_{1,\theta}\mathbf{d}_{1,\theta}^H \left|\mathcal{B}_{2,\theta}(\mathbf{h}_2)\right|^2 \sin\theta d\theta \qquad (3.223)\\ &= \left(\mathbf{I}_{M_0} \otimes \mathbf{h}_2\right)^H \mathbf{\Gamma}_{\mathrm{f}} \left(\mathbf{I}_{M_0} \otimes \mathbf{h}_2\right),\\ \mathbf{\Gamma}_{\mathrm{b},\mathbf{h}_2} &= \int_{\pi/2}^{\pi} \mathbf{d}_{1,\theta}\mathbf{d}_{1,\theta}^H \left|\mathcal{B}_{2,\theta}(\mathbf{h}_2)\right|^2 \sin\theta d\theta \qquad (3.224)\\ &= \left(\mathbf{I}_{M_0} \otimes \mathbf{h}_2\right)^H \mathbf{\Gamma}_{\mathrm{b}} \left(\mathbf{I}_{M_0} \otimes \mathbf{h}_2\right).\end{aligned}$$

In the same way, when $\mathbf{h}_1$ is fixed and given, and thanks to (2.40), we can write the FBR as

$$\mathcal{F}(\mathbf{h}_2|\mathbf{h}_1) = \frac{\mathbf{h}_2^H \mathbf{\Gamma}_{\mathrm{f},\mathbf{h}_1} \mathbf{h}_2}{\mathbf{h}_2^H \mathbf{\Gamma}_{\mathrm{b},\mathbf{h}_1} \mathbf{h}_2}, \tag{3.225}$$

where

$$\begin{aligned}\mathbf{\Gamma}_{\mathrm{f},\mathbf{h}_1} &= \int_0^{\pi/2} \mathbf{d}_{2,\theta}\mathbf{d}_{2,\theta}^H \left|\mathcal{B}_{1,\theta}(\mathbf{h}_1)\right|^2 \sin\theta d\theta \qquad (3.226)\\ &= \left(\mathbf{h}_1 \otimes \mathbf{I}_{M_0}\right)^H \mathbf{\Gamma}_{\mathrm{f}} \left(\mathbf{h}_1 \otimes \mathbf{I}_{M_0}\right),\\ \mathbf{\Gamma}_{\mathrm{b},\mathbf{h}_1} &= \int_{\pi/2}^{\pi} \mathbf{d}_{2,\theta}\mathbf{d}_{2,\theta}^H \left|\mathcal{B}_{1,\theta}(\mathbf{h}_1)\right|^2 \sin\theta d\theta \qquad (3.227)\\ &= \left(\mathbf{h}_1 \otimes \mathbf{I}_{M_0}\right)^H \mathbf{\Gamma}_{\mathrm{b}} \left(\mathbf{h}_1 \otimes \mathbf{I}_{M_0}\right).\end{aligned}$$

3.4.2 Cardioid

The $(M_0 - 1)$th-order cardioid has a unique null of multiplicity $M_0 - 1$ in the direction π. Therefore, the ith derivative, with $i = 0, 1, \ldots, M_0 - 2$, of the beampattern of $\mathbf{h}_2$ with respect to $\cos\theta$ is equal to 0 at $\cos\pi = -1$, i.e.,

$$\left.\frac{d^i \mathcal{B}_{2,\theta}(\mathbf{h}_2)}{d\cos^i\theta}\right|_{\cos\theta=-1} = \mathcal{B}_{2,\pi}^{[i]}(\mathbf{h}_2) = 0, \tag{3.228}$$

with

$$\mathcal{B}_{2,\pi}^{[0]}\left(\mathbf{h}_2\right) = \mathcal{B}_{2,\pi}\left(\mathbf{h}_2\right).$$

We easily find that

$$\mathcal{B}_{2,\pi}^{[i]}\left(\mathbf{h}_2\right) = \left(\jmath\omega\delta/c\right)^i \left(\mathbf{\Sigma}^i \mathbf{d}_{2,\pi}\right)^H \mathbf{h}_2, \tag{3.229}$$

where

$$\mathbf{\Sigma} = \text{diag}\left(0, 1, \ldots, M_0 - 1\right) \tag{3.230}$$

is a diagonal matrix of size $M_0 \times M_0$. Combining the distortionless constraint, i.e.,

$$\mathcal{B}_{2,0}\left(\mathbf{h}_2\right) = \mathbf{d}_{2,0}^H \mathbf{h}_2 = 1, \tag{3.231}$$

with the $M_0 - 1$ equations from (3.228), we obtain a linear system of M_0 equations with M_0 unknowns:

$$\mathbf{D}_{2,\pi}^H \mathbf{h}_2 = \mathbf{i}, \tag{3.232}$$

where

$$\mathbf{D}_{2,\pi}^H = \begin{bmatrix} \mathbf{d}_{2,0}^H \\ \left(\mathbf{\Sigma}^0 \mathbf{d}_{2,\pi}\right)^H \\ \left(\mathbf{\Sigma}^1 \mathbf{d}_{2,\pi}\right)^H \\ \vdots \\ \left(\mathbf{\Sigma}^{M_0-2} \mathbf{d}_{2,\pi}\right)^H \end{bmatrix} \tag{3.233}$$

and $\mathbf{i}$ is the first column of $\mathbf{I}_{M_0}$. Therefore, the cardioid of order $M_0 - 1$ at the second ULA is

$$\mathbf{h}_{2,\text{C}} = \mathbf{D}_{2,\pi}^{-H} \mathbf{i}. \tag{3.234}$$

For the first filter, we take the DS beamformer, i.e.,

$$\mathbf{h}_{1,\text{DS}} = \frac{\mathbf{d}_{1,0}}{M_0}, \tag{3.235}$$

which maximizes the WNG. As a result, the robust global cardioid of order, at least, $M_0 - 1$ is

$$\mathbf{h}_{\text{C}} = \mathbf{h}_{1,\text{DS}} \otimes \mathbf{h}_{2,\text{C}}. \tag{3.236}$$

Figure 3.19 displays the directivity patterns of the robust global cardioid, $\mathbf{h}_{\text{C}}$, for $f = 1$ kHz, $\delta = 1$ cm, and different numbers of sensors M. Figure 3.20 shows plots of the DFs and WNGs of the robust global cardioid, $\mathbf{h}_{\text{C}}$, as a

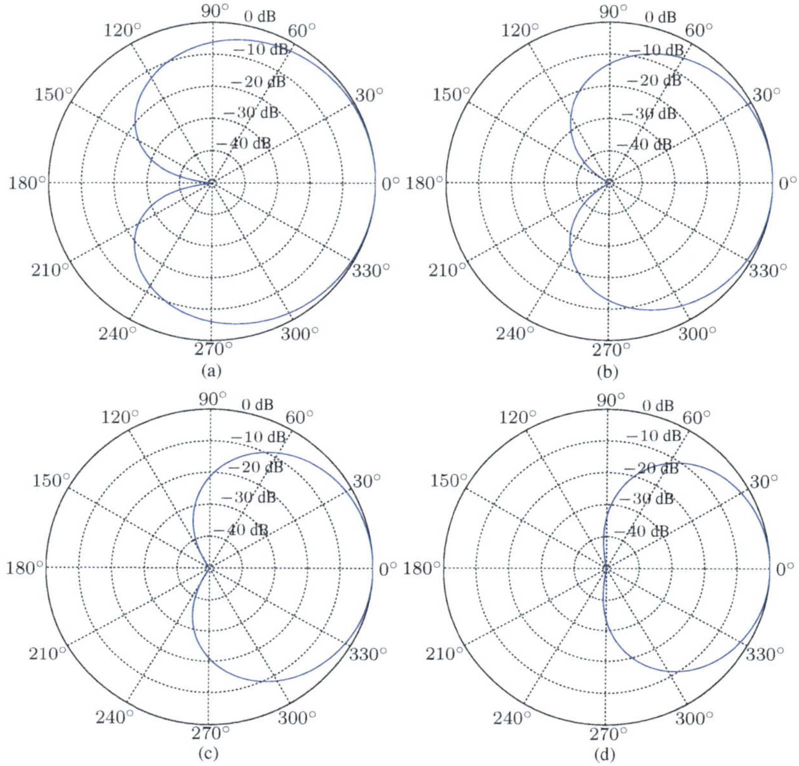

Fig. 3.19 Beampatterns of the robust global cardioid, $\mathbf{h}_C$, for $f = 1$ kHz, $\delta = 1$ cm, and different numbers of sensors $M = M_0^2$: (a) $M_0 = 2$, (b) $M_0 = 3$, (c) $M_0 = 4$, and (d) $M_0 = 5$.

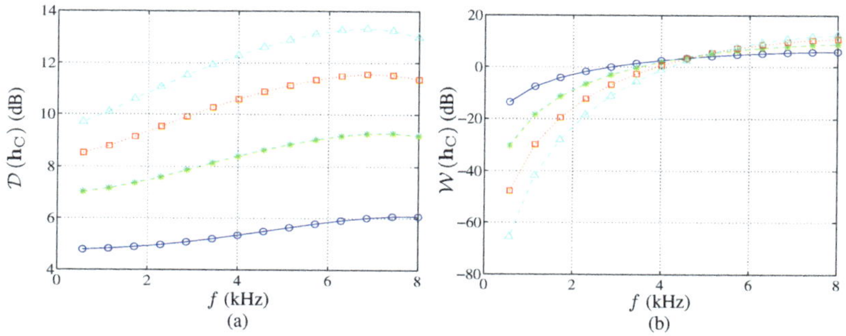

Fig. 3.20 Performance of the robust global cardioid, $\mathbf{h}_C$, as a function of frequency for $\delta = 1$ cm, and different numbers of sensors $M = M_0^2$: $M_0 = 2$ (solid line with circles), $M_0 = 3$ (dashed line with asterisks), $M_0 = 4$ (dotted line with squares), and $M_0 = 5$ (dash-dot line with triangles). (a) DF and (b) WNG.

function of frequency. We observe that the DF increases as we increase the numbers of sensors, but the WNG decreases at low frequencies.

To get an even more robust beamformer, we may optimize

$$\min_{\mathbf{h}_2} \mathbf{h}_2^H \left(\mathbf{D}'_{2,\pi}\mathbf{D}'^H_{2,\pi} + \epsilon_2 \mathbf{I}_{M_0}\right)\mathbf{h}_2 \quad \text{subject to} \quad \mathbf{C}_{2,\pi}^H \mathbf{h}_2 = \mathbf{i}_{\mathrm{c}}, \tag{3.237}$$

where

$$\mathbf{D}'_{2,\pi} = \left[\boldsymbol{\Sigma}^1 \mathbf{d}_{2,\pi} \ \boldsymbol{\Sigma}^2 \mathbf{d}_{2,\pi} \ \cdots \ \boldsymbol{\Sigma}^{M_0-2}\mathbf{d}_{2,\pi}\right] \tag{3.238}$$

is a matrix of size $M_0 \times (M_0 - 2)$, $\epsilon_2 > 0$ is the regularization parameter, and

$$\mathbf{C}_{2,\pi}^H = \begin{bmatrix} \mathbf{d}_{2,0}^H \\ \mathbf{d}_{2,\pi}^H \end{bmatrix} \tag{3.239}$$

is the constraint matrix of size $2 \times M_0$. We see that with $\mathbf{C}_{2,\pi}$, the two main constraints are fulfilled, i.e., the distortionless one and a null in the direction π. We find that the optimal filter is

$$\begin{aligned} \mathbf{h}_{2,\mathrm{C},\epsilon_2} &= \left(\mathbf{D}'_{2,\pi}\mathbf{D}'^H_{2,\pi} + \epsilon_2 \mathbf{I}_{M_0}\right)^{-1} \mathbf{C}_{2,\pi} \\ &\times \left[\mathbf{C}_{2,\pi}^H \left(\mathbf{D}'_{2,\pi}\mathbf{D}'^H_{2,\pi} + \epsilon_2 \mathbf{I}_{M_0}\right)^{-1} \mathbf{C}_{2,\pi}\right]^{-1} \mathbf{i}_{\mathrm{c}}. \end{aligned} \tag{3.240}$$

Therefore, the more robust global cardioid is

$$\mathbf{h}_{\mathrm{C},\epsilon_2} = \mathbf{h}_{1,\mathrm{DS}} \otimes \mathbf{h}_{2,\mathrm{C},\epsilon_2}. \tag{3.241}$$

Figure 3.21 displays the directivity patterns of the robust global cardioid, $\mathbf{h}_{\mathrm{C},\epsilon_2}$, for $f = 1$ kHz, $\delta = 1$ cm, $M_0 = 4$, and several values of ϵ_2. Figure 3.22 shows plots of the DFs and WNGs of the robust global cardioid, $\mathbf{h}_{\mathrm{C},\epsilon_2}$, as a function of frequency. We observe that the WNG increases as we increase ϵ_2, but the DF decreases.

3.4.3 Dipole

The design of the global dipole is a bit different from the design of the global cardioid as explained below.

The dipole of order $M_0 - 1$ has also a unique null of multiplicity $M_0 - 1$ but in the direction $\pi/2$. Since we have a null with maximum multiplicity, the ith ($i = 0, 1, \ldots, M_0 - 2$) derivative of the beampattern of $\mathbf{h}_2$ with respect to $\cos\theta$ is equal to 0 at $\cos(\pi/2) = 0$, i.e.,

$$\left.\frac{d^i \mathcal{B}_{2,\theta}(\mathbf{h}_2)}{d\cos^i\theta}\right|_{\cos\theta=0} = \mathcal{B}_{2,\pi/2}^{[i]}(\mathbf{h}_2) = 0, \tag{3.242}$$

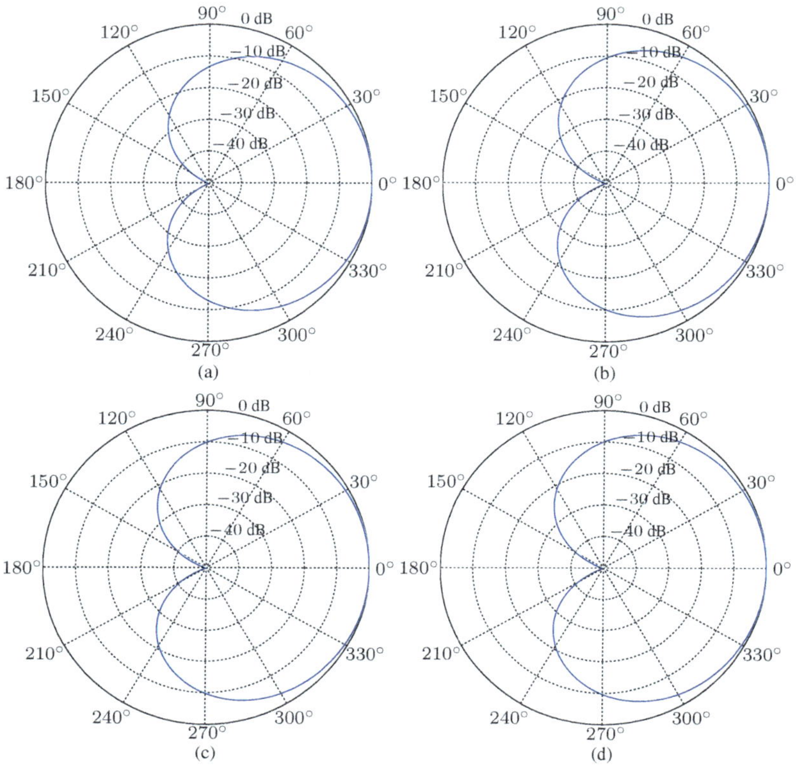

Fig. 3.21 Beampatterns of the robust global cardioid, $\mathbf{h}_{\mathrm{C},\epsilon_2}$, for $f = 1$ kHz, $\delta = 1$ cm, $M_0 = 4$, and several values of ϵ_2: (a) $\epsilon_2 = 0.001$, (b) $\epsilon_2 = 0.01$, (c) $\epsilon_2 = 0.1$, and (d) $\epsilon_2 = 1$.

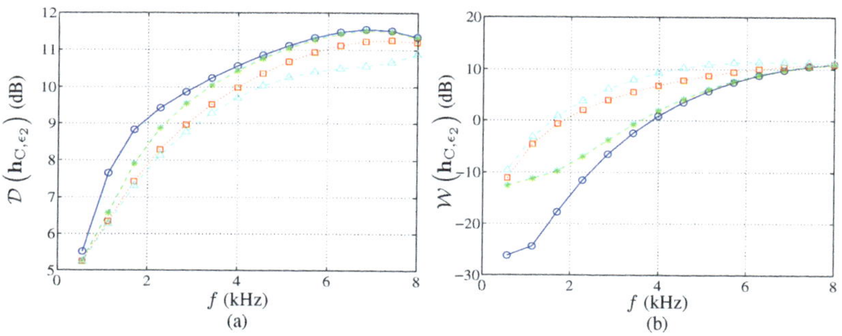

Fig. 3.22 Performance of the robust global cardioid, $\mathbf{h}_{\mathrm{C},\epsilon_2}$, as a function of frequency for $\delta = 1$ cm, $M_0 = 4$, and several values of ϵ_2: $\epsilon_2 = 0.001$ (solid line with circles), $\epsilon_2 = 0.01$ (dashed line with asterisks), $\epsilon_2 = 0.1$ (dotted line with squares), and $\epsilon_2 = 1$ (dash-dot line with triangles). (a) DF and (b) WNG.

with

$$\mathcal{B}^{[0]}_{2,\pi/2}(\mathbf{h}_2) = \mathcal{B}_{2,\pi/2}(\mathbf{h}_2).$$

We get

$$\mathcal{B}^{[i]}_{2,\pi/2}(\mathbf{h}_2) = (\jmath\omega\delta/c)^i \left(\mathbf{\Sigma}^i \mathbf{d}_{2,\pi/2}\right)^H \mathbf{h}_2, \tag{3.243}$$

where $\mathbf{\Sigma}$ is defined in (3.230). Combining the distortionless constraint with the $M_0 - 1$ constraints from (3.242), we have

$$\mathbf{D}^H_{2,\pi/2}\mathbf{h}_2 = \mathbf{i}, \tag{3.244}$$

where

$$\mathbf{D}^H_{2,\pi/2} = \begin{bmatrix} \mathbf{d}^H_{2,0} \\ \left(\mathbf{\Sigma}^0 \mathbf{d}_{2,\pi/2}\right)^H \\ \left(\mathbf{\Sigma}^1 \mathbf{d}_{2,\pi/2}\right)^H \\ \vdots \\ \left(\mathbf{\Sigma}^{M_0-2} \mathbf{d}_{2,\pi/2}\right)^H \end{bmatrix}. \tag{3.245}$$

As a result, the dipole of order $M_0 - 1$ at the second ULA is

$$\mathbf{h}_{2,\mathrm{D}} = \mathbf{D}^{-H}_{2,\pi/2}\mathbf{i}. \tag{3.246}$$

Another feature of the dipole is that it has a 1 in the direction π. To ensure that the global beampattern has also a 1 at π, we must add this constraint in the design of the first filter. We deduce that the constraint equation for $\mathbf{h}_1$ is

$$\mathbf{C}^H_{1,\pi}\mathbf{h}_1 = \begin{bmatrix} 1 \\ 1 \end{bmatrix}, \tag{3.247}$$

where

$$\mathbf{C}^H_{1,\pi} = \begin{bmatrix} \mathbf{d}^H_{1,0} \\ \mathbf{d}^H_{1,\pi} \end{bmatrix} \tag{3.248}$$

is the constraint matrix of size $2 \times M_0$. Since we want to maximize the WNG of $\mathbf{h}_1$ subject to (3.247), we find the minimum-norm beamformer:

$$\mathbf{h}_{1,\mathrm{MN}} = \mathbf{C}_{1,\pi}\left(\mathbf{C}^H_{1,\pi}\mathbf{C}_{1,\pi}\right)^{-1} \begin{bmatrix} 1 \\ 1 \end{bmatrix}. \tag{3.249}$$

Now, that the two filters are derived, we deduce that the robust global dipole of order, at least, $M_0 - 1$ is

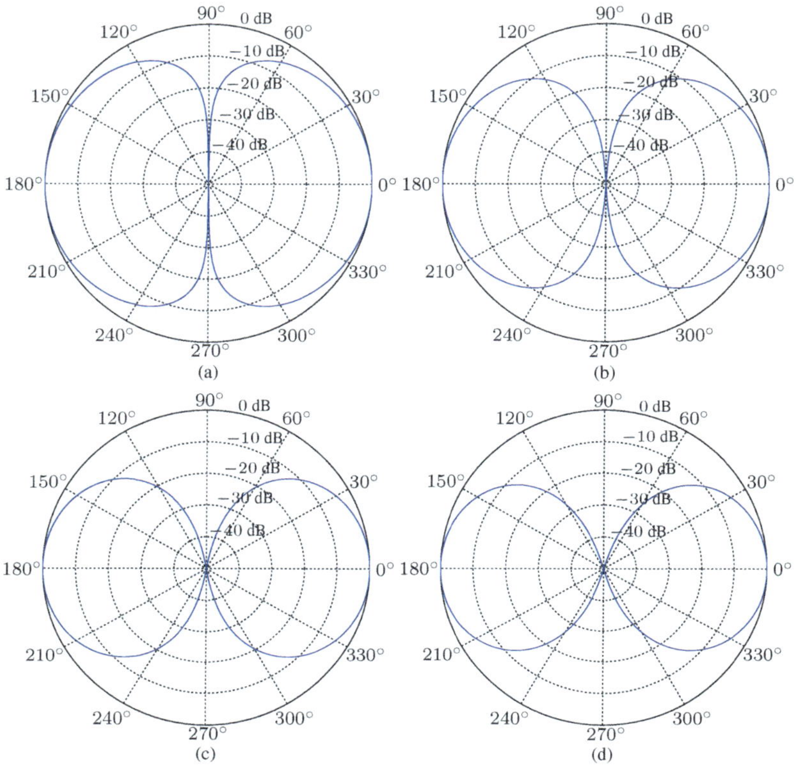

Fig. 3.23 Beampatterns of the robust global dipole, $\mathbf{h}_\mathrm{D}$, for $f = 1$ kHz, $\delta = 1$ cm, and different numbers of sensors $M = M_0^2$: (a) $M_0 = 2$, (b) $M_0 = 3$, (c) $M_0 = 4$, and (d) $M_0 = 5$.

$$\mathbf{h}_\mathrm{D} = \mathbf{h}_{1,\mathrm{MN}} \otimes \mathbf{h}_{2,\mathrm{D}}. \tag{3.250}$$

Figure 3.23 displays the directivity patterns of the robust global dipole, $\mathbf{h}_\mathrm{D}$, for $f = 1$ kHz, $\delta = 1$ cm, and different numbers of sensors M. Figure 3.24 shows plots of the DFs and WNGs of the robust global dipole, $\mathbf{h}_\mathrm{D}$, as a function of frequency. We observe that the DF generally increases as we increase the numbers of sensors, but the WNG decreases.

As we did for the cardioid, we can derive a more robust global dipole:

$$\mathbf{h}_{\mathrm{D},\epsilon_2} = \mathbf{h}_{1,\mathrm{MN}} \otimes \mathbf{h}_{2,\mathrm{D},\epsilon_2}, \tag{3.251}$$

where

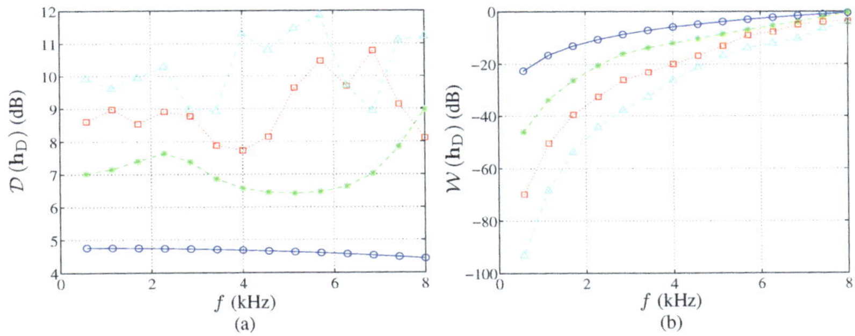

Fig. 3.24 Performance of the robust global dipole, $\mathbf{h}_D$, as a function of frequency for $\delta = 1$ cm and different numbers of sensors $M = M_0^2$: $M_0 = 2$ (solid line with circles), $M_0 = 3$ (dashed line with asterisks), $M_0 = 4$ (dotted line with squares), and $M_0 = 5$ (dash-dot line with triangles). (a) DF and (b) WNG.

$$\mathbf{h}_{2,D,\epsilon_2} = \left(\mathbf{D}'_{2,\pi/2}\mathbf{D}'^H_{2,\pi/2} + \epsilon_2 \mathbf{I}_{M_0}\right)^{-1} \mathbf{C}_{2,\pi/2} \times \left[\mathbf{C}^H_{2,\pi/2}\left(\mathbf{D}'_{2,\pi/2}\mathbf{D}'^H_{2,\pi/2} + \epsilon_2 \mathbf{I}_{M_0}\right)^{-1} \mathbf{C}_{2,\pi/2}\right]^{-1} \mathbf{i}_c, \tag{3.252}$$

with

$$\mathbf{D}'_{2,\pi/2} = \left[\, \boldsymbol{\Sigma}^1 \mathbf{d}_{2,\pi/2}\ \boldsymbol{\Sigma}^2 \mathbf{d}_{2,\pi/2}\ \cdots\ \boldsymbol{\Sigma}^{M_0-2} \mathbf{d}_{2,\pi/2} \,\right] \tag{3.253}$$

and

$$\mathbf{C}^H_{2,\pi/2} = \begin{bmatrix} \mathbf{d}^H_{2,0} \\ \mathbf{d}^H_{2,\pi/2} \end{bmatrix}. \tag{3.254}$$

Figure 3.25 displays the directivity patterns of the robust global dipole, $\mathbf{h}_{D,\epsilon_2}$, for $f = 1$ kHz, $\delta = 1$ cm, $M_0 = 4$, and several values of ϵ_2. Figure 3.26 shows plots of the DFs and WNGs of the robust global dipole, $\mathbf{h}_{D,\epsilon_2}$, as a function of frequency. We observe that the WNG increases as we increase ϵ_2.

3.4.4 Hypercardioid

The hypercardioid is usually obtained from the maximization of the DF.

For the first filter, $\mathbf{h}_1$, we take the DS beamformer, i.e., $\mathbf{h}_1 = \mathbf{h}_{1,DS}$, so that its WNG is maximized. By definition, the DF of $\mathbf{h}_2$ is

$$\mathcal{D}(\mathbf{h}_2) = \frac{\left|\mathbf{h}_2^H \mathbf{d}_{2,0}\right|^2}{\mathbf{h}_2^H \boldsymbol{\Gamma}_2 \mathbf{h}_2}. \tag{3.255}$$

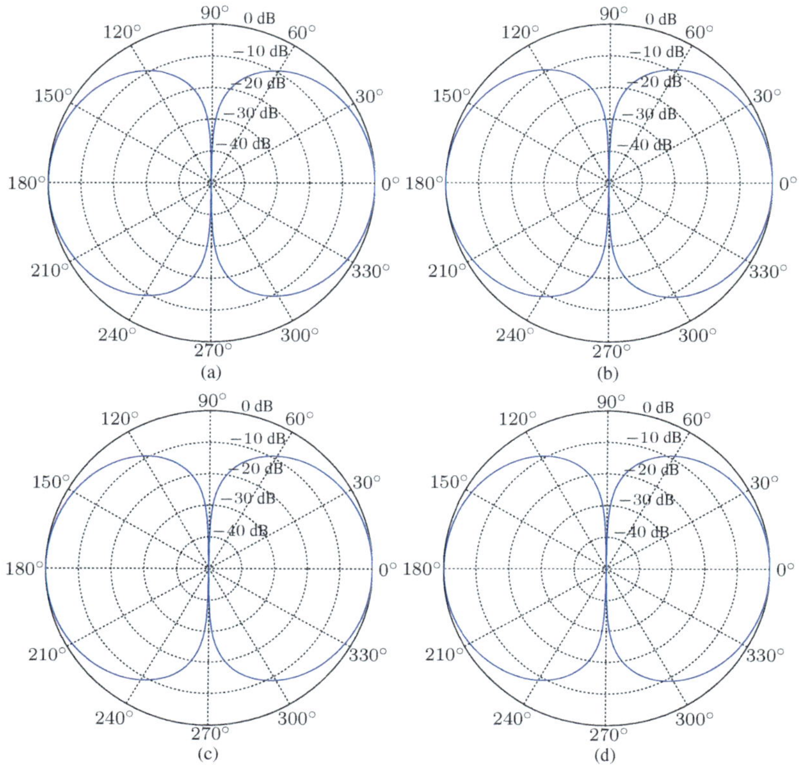

Fig. 3.25 Beampatterns of the robust global dipole, $\mathbf{h}_{\mathrm{D},\epsilon_2}$, for $f = 1$ kHz, $\delta = 1$ cm, $M_0 = 4$, and several values of ϵ_2: (a) $\epsilon_2 = 0.001$, (b) $\epsilon_2 = 0.01$, (c) $\epsilon_2 = 0.1$, and (d) $\epsilon_2 = 1$.

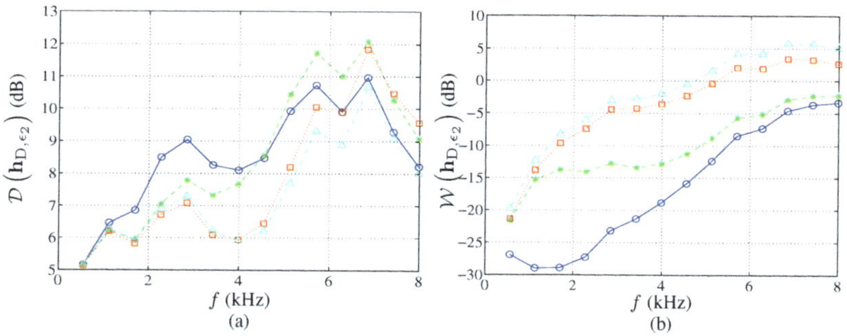

Fig. 3.26 Performance of the robust global dipole, $\mathbf{h}_{\mathrm{D},\epsilon_2}$, as a function of frequency for $\delta = 1$ cm, $M_0 = 4$, and several values of ϵ_2: $\epsilon_2 = 0.001$ (solid line with circles), $\epsilon_2 = 0.01$ (dashed line with asterisks), $\epsilon_2 = 0.1$ (dotted line with squares), and $\epsilon_2 = 1$ (dash-dot line with triangles). (a) DF and (b) WNG.

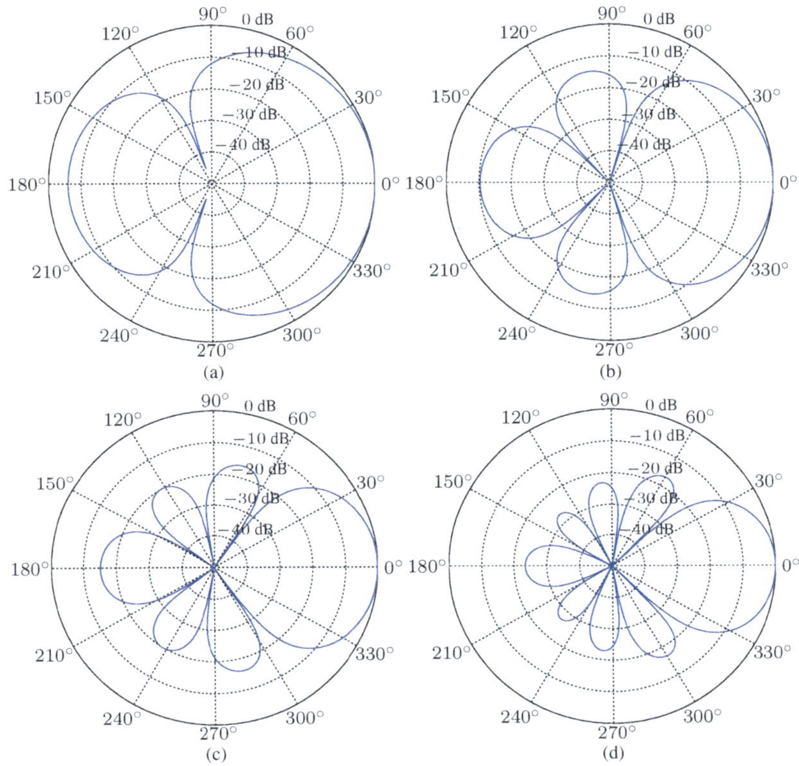

Fig. 3.27 Beampatterns of the robust global hypercardioid, $\mathbf{h}_{\mathrm{H}}$, for $f = 1$ kHz, $\delta = 5$ mm, and different numbers of sensors $M = M_0^2$: (a) $M_0 = 2$, (b) $M_0 = 3$, (c) $M_0 = 4$, and (d) $M_0 = 5$.

The maximization of $\mathcal{D}(\mathbf{h}_2)$ gives the hypercardioid of order $M_0 - 1$ at the second ULA:

$$\mathbf{h}_{2,\mathrm{H}} = \frac{\mathbf{\Gamma}_2^{-1}\mathbf{d}_{2,0}}{\mathbf{d}_{2,0}^H \mathbf{\Gamma}_2^{-1}\mathbf{d}_{2,0}}. \tag{3.256}$$

As a result, the robust global hypercardioid of order, at least, $M_0 - 1$ is

$$\mathbf{h}_{\mathrm{H}} = \mathbf{h}_{1,\mathrm{DS}} \otimes \mathbf{h}_{2,\mathrm{H}}. \tag{3.257}$$

Figure 3.27 displays the directivity patterns of the robust global hypercardioid, $\mathbf{h}_{\mathrm{H}}$, for $f = 1$ kHz, $\delta = 5$ mm, and different numbers of sensors M. Figure 3.28 shows plots of the DFs and WNGs of the robust global hypercardioid, $\mathbf{h}_{\mathrm{H}}$, as a function of frequency. We observe that the DF generally increases as we increase the numbers of sensors, but the WNG decreases.

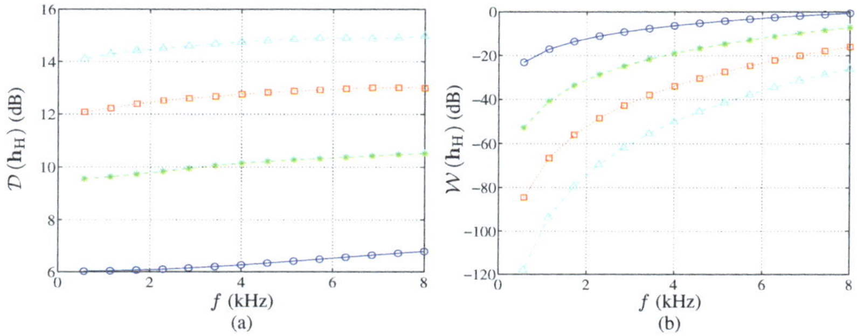

Fig. 3.28 Performance of the robust global hypercardioid, $\mathbf{h}_{\rm H}$, as a function of frequency for $\delta = 5$ mm and different numbers of sensors $M = M_0^2$: $M_0 = 2$ (solid line with circles), $M_0 = 3$ (dashed line with asterisks), $M_0 = 4$ (dotted line with squares), and $M_0 = 5$ (dash-dot line with triangles). (a) DF and (b) WNG.

To make this beamformer even more robust, we can use the following instead

$$\mathbf{h}_{{\rm H},\epsilon_2} = \mathbf{h}_{1,{\rm DS}} \otimes \mathbf{h}_{2,{\rm H},\epsilon_2}, \tag{3.258}$$

where

$$\mathbf{h}_{2,{\rm H},\epsilon_2} = \frac{\left(\mathbf{\Gamma}_2 + \epsilon_2 \mathbf{I}_{M_0}\right)^{-1} \mathbf{d}_{2,0}}{\mathbf{d}_{2,0}^H \left(\mathbf{\Gamma}_2 + \epsilon_2 \mathbf{I}_{M_0}\right)^{-1} \mathbf{d}_{2,0}}, \tag{3.259}$$

with $\epsilon_2 \geq 0$ being the regularization parameter.

Other possibilities can be borrowed from Subsection 3.1.2 by taking $\theta_{\rm d} = 0$.

Figure 3.29 displays the directivity patterns of the robust global dipole, $\mathbf{h}_{{\rm D},\epsilon_2}$, for $f = 1$ kHz, $\delta = 1$ cm, $M_0 = 4$, and several values of ϵ_2. Figure 3.30 shows plots of the DFs and WNGs of the robust global dipole, $\mathbf{h}_{{\rm D},\epsilon_2}$, as a function of frequency. We observe that the WNG increases as we increase ϵ_2, but the DF decreases.

3.4.5 Supercardioid

As we should expect, different versions of the supercardioid can be derived.

In the first version, we choose the DS beamformer for the first filter, i.e., $\mathbf{h}_1 = \mathbf{h}_{1,{\rm DS}}$. By definition, the FBR of $\mathbf{h}_2$ is given in (3.213) and we want to maximize this quantity. Let $\mathbf{t}_2$ be the eigenvector corresponding to the maximum eigenvalue of the matrix $\mathbf{\Gamma}_{{\rm b},2}^{-1}\mathbf{\Gamma}_{{\rm f},2}$. It is clear that the filter $\mathbf{h}_2 = \alpha \mathbf{t}_2$, where $\alpha \neq 0$ is an arbitrary complex number, maximizes this FBR. Taking the distortionless constraint into account, we deduce that the supercardioid of order $M_0 - 1$ at the second ULA is

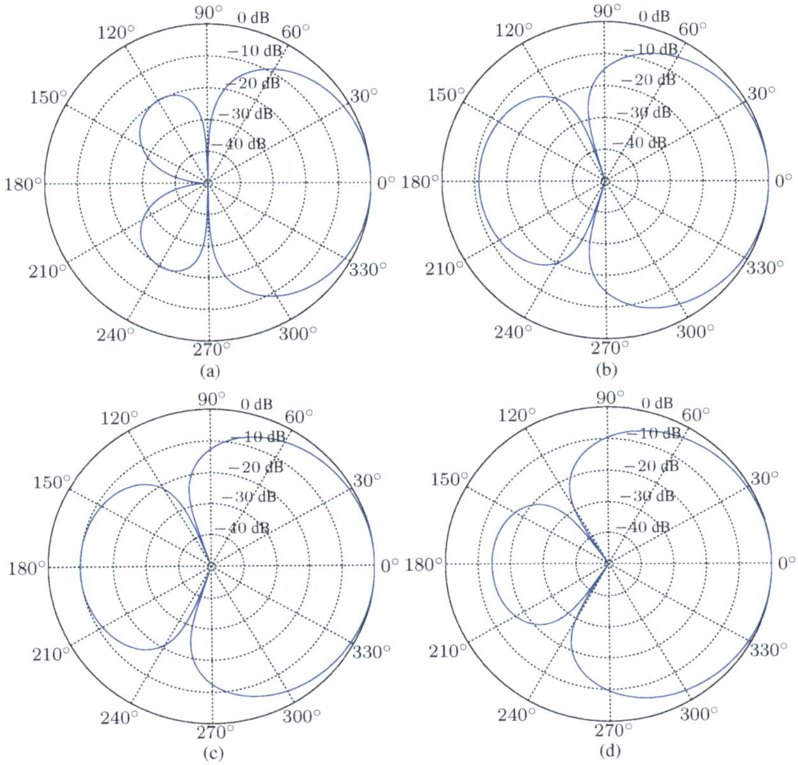

Fig. 3.29 Beampatterns of the robust global hypercardioid, $\mathbf{h}_{\mathrm{H},\epsilon_2}$, for $f = 1$ kHz, $\delta = 5$ mm, $M_0 = 4$, and several values of ϵ_2: (a) $\epsilon_2 = 10^{-5}$, (b) $\epsilon_2 = 10^{-4}$, (c) $\epsilon_2 = 10^{-3}$, and (d) $\epsilon_2 = 10^{-2}$.

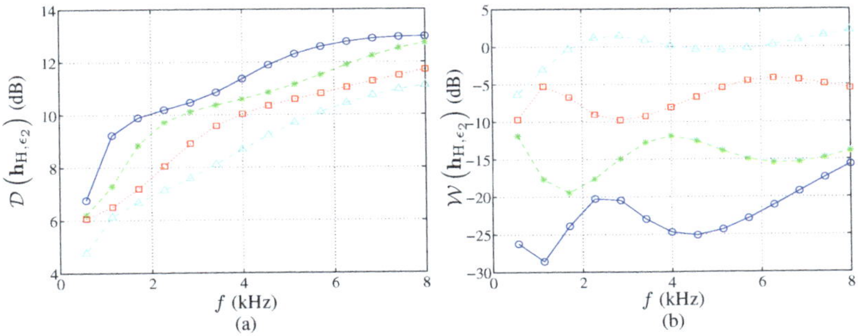

Fig. 3.30 Performance of the robust global hypercardioid, $\mathbf{h}_{\mathrm{H},\epsilon_2}$, as a function of frequency for $\delta = 5$ mm, $M_0 = 4$, and several values of ϵ_2: $\epsilon_2 = 10^{-5}$ (solid line with circles), $\epsilon_2 = 10^{-4}$ (dashed line with asterisks), $\epsilon_2 = 10^{-3}$ (dotted line with squares), and $\epsilon_2 = 10^{-2}$ (dash-dot line with triangles). (a) DF and (b) WNG.

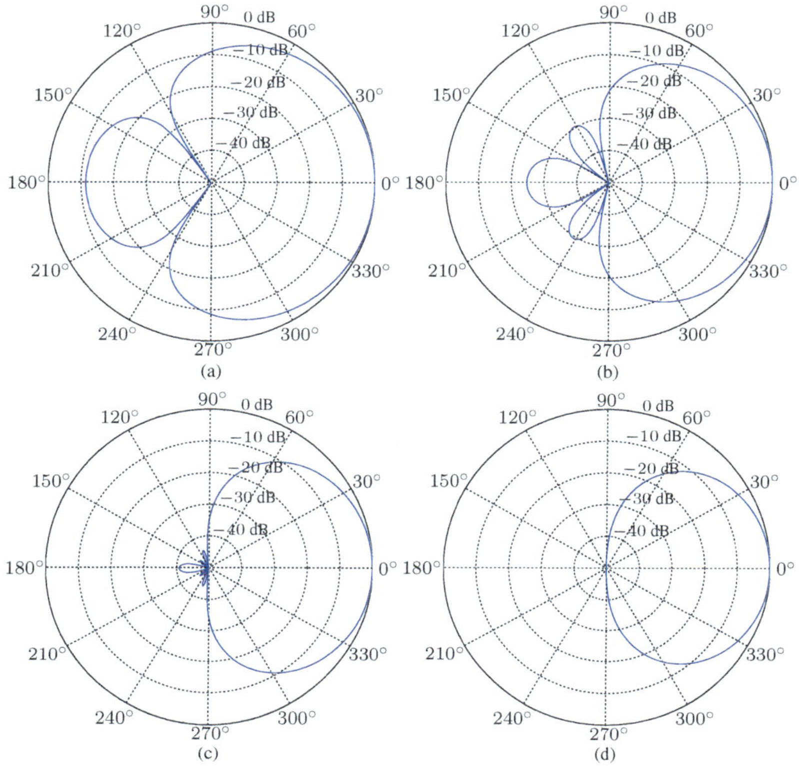

Fig. 3.31 Beampatterns of the robust global supercardioid, $\mathbf{h}_{\mathrm{S}1}$, for $f = 1$ kHz, $\delta = 5$ mm, and different numbers of sensors $M = M_0^2$: (a) $M_0 = 2$, (b) $M_0 = 3$, (c) $M_0 = 4$, and (d) $M_0 = 5$.

$$\mathbf{h}_{2,\mathrm{S}} = \frac{\mathbf{t}_2}{\mathbf{d}_{2,0}^H \mathbf{t}_2}. \tag{3.260}$$

Therefore, the robust global supercardioid of order, at least, $M_0 - 1$ is

$$\mathbf{h}_{\mathrm{S}1} = \mathbf{h}_{1,\mathrm{DS}} \otimes \mathbf{h}_{2,\mathrm{S}}. \tag{3.261}$$

Figure 3.31 displays the directivity patterns of the robust global supercardioid, $\mathbf{h}_{\mathrm{S}1}$, for $f = 1$ kHz, $\delta = 5$ mm, and different numbers of sensors M. Figure 3.32 shows plots of the DFs, WNGs, and FBRs of the robust global supercardioid, $\mathbf{h}_{\mathrm{S}1}$, as a function of frequency. We observe that the DF and FBR increase as we increase the number of sensors, but the WNG decreases.

To have a robust version of the supercardioid, $\mathbf{h}_{2,\mathrm{S}}$, we need now to consider the matrix $\left(\mathbf{\Gamma}_{\mathrm{b},2} + \epsilon_2 \mathbf{I}_{M_0}\right)^{-1} \mathbf{\Gamma}_{\mathrm{f},2}$. By taking the eigenvector corresponding to the maximum eigenvalue of this matrix that we denote $\mathbf{t}_{2,\epsilon_2}$, we find that the robust supercardioid at the second ULA is

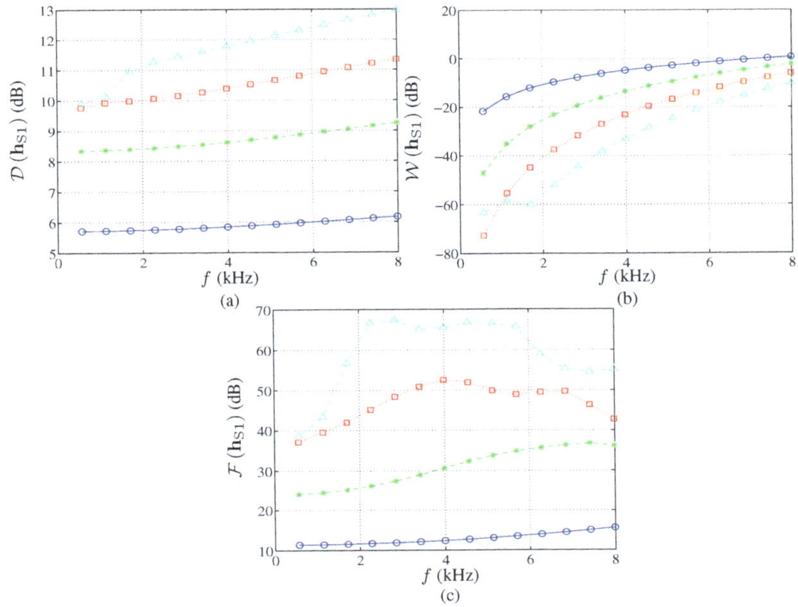

Fig. 3.32 Performance of the robust global supercardioid, $\mathbf{h}_{S1}$, as a function of frequency for $\delta = 5$ mm and different numbers of sensors $M = M_0^2$: $M_0 = 2$ (solid line with circles), $M_0 = 3$ (dashed line with asterisks), $M_0 = 4$ (dotted line with squares), and $M_0 = 5$ (dash-dot line with triangles). (a) DF, (b) WNG, and (c) FBR.

$$\mathbf{h}_{2,\mathrm{S},\epsilon_2} = \frac{\mathbf{t}_{2,\epsilon_2}}{\mathbf{d}_{2,0}^H \mathbf{t}_{2,\epsilon_2}}. \tag{3.262}$$

Then, the more robust global supercardioid is

$$\mathbf{h}_{\mathrm{S1},\epsilon_2} = \mathbf{h}_{1,\mathrm{DS}} \otimes \mathbf{h}_{2,\mathrm{S},\epsilon_2}. \tag{3.263}$$

Figure 3.33 displays the directivity patterns of the robust global supercardioid, $\mathbf{h}_{\mathrm{S1},\epsilon_2}$, for $f = 1$ kHz, $\delta = 5$ mm, $M_0 = 4$, and several values of ϵ_2. Figure 3.34 shows plots of the DFs, WNGs, and FBRs of the robust global supercardioid, $\mathbf{h}_{\mathrm{S1},\epsilon_2}$, as a function of frequency. We observe that the WNG increases as we increase ϵ_2, but the DF and FBR decrease.

Assume that $\mathbf{h}_1 = \mathbf{h}_{1,\mathrm{DS}}$. Substituting this filter into (3.225), we get

$$\mathcal{F}\left(\mathbf{h}_2 | \mathbf{h}_{1,\mathrm{DS}}\right) = \frac{\mathbf{h}_2^H \mathbf{\Gamma}_{\mathrm{f},\mathbf{h}_{1,\mathrm{DS}}} \mathbf{h}_2}{\mathbf{h}_2^H \mathbf{\Gamma}_{\mathrm{b},\mathbf{h}_{1,\mathrm{DS}}} \mathbf{h}_2}, \tag{3.264}$$

where

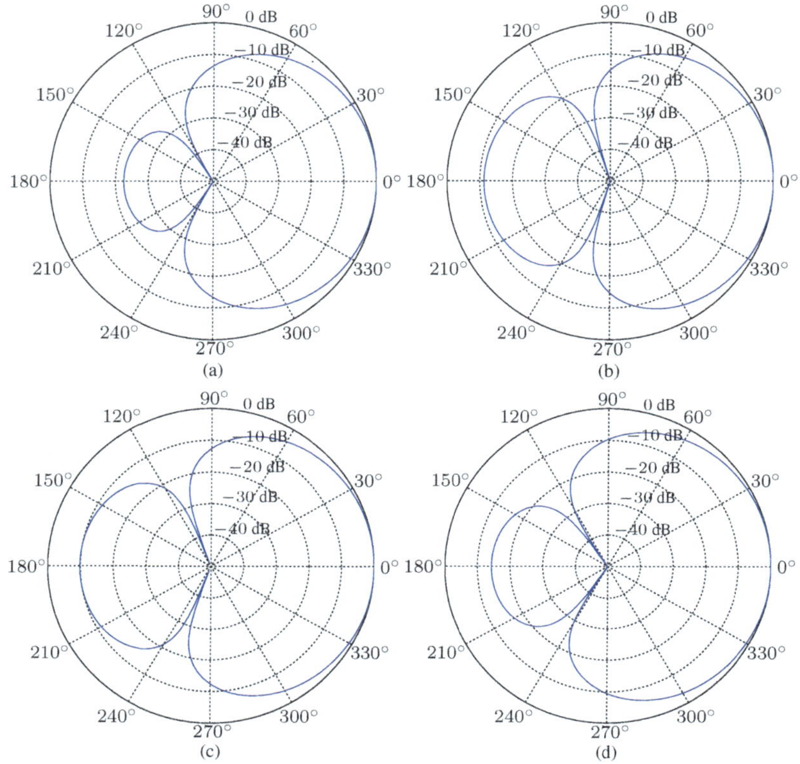

Fig. 3.33 Beampatterns of the robust global supercardioid, $\mathbf{h}_{\mathrm{S1},\epsilon_2}$, for $f = 1$ kHz, $\delta = 5$ mm, $M_0 = 4$, and several values of ϵ_2: (a) $\epsilon_2 = 10^{-5}$, (b) $\epsilon_2 = 10^{-4}$, (c) $\epsilon_2 = 10^{-3}$, and (d) $\epsilon_2 = 10^{-2}$.

$$\mathbf{\Gamma}_{\mathrm{f},\mathbf{h}_{1,\mathrm{DS}}} = \left(\mathbf{h}_{1,\mathrm{DS}} \otimes \mathbf{I}_{M_0}\right)^H \mathbf{\Gamma}_{\mathrm{f}} \left(\mathbf{h}_{1,\mathrm{DS}} \otimes \mathbf{I}_{M_0}\right), \tag{3.265}$$

$$\mathbf{\Gamma}_{\mathrm{b},\mathbf{h}_{1,\mathrm{DS}}} = \left(\mathbf{h}_{1,\mathrm{DS}} \otimes \mathbf{I}_{M_0}\right)^H \mathbf{\Gamma}_{\mathrm{b}} \left(\mathbf{h}_{1,\mathrm{DS}} \otimes \mathbf{I}_{M_0}\right). \tag{3.266}$$

Let $\mathbf{t}_{2,\mathbf{h}_{1,\mathrm{DS}}}$ be the eigenvector associated with the maximum eigenvalue of $\mathbf{\Gamma}^{-1}_{\mathrm{b},\mathbf{h}_{1,\mathrm{DS}}} \mathbf{\Gamma}_{\mathrm{f},\mathbf{h}_{1,\mathrm{DS}}}$. Then, it is clear that another supercardioid at the second ULA is

$$\mathbf{h}_{2,\mathrm{S2}} = \frac{\mathbf{t}_{2,\mathbf{h}_{1,\mathrm{DS}}}}{\mathbf{d}_{2,0}^H \mathbf{t}_{2,\mathbf{h}_{1,\mathrm{DS}}}}. \tag{3.267}$$

Therefore, another version of the robust global supercardioid is

$$\mathbf{h}_{\mathrm{S2}} = \mathbf{h}_{1,\mathrm{DS}} \otimes \mathbf{h}_{2,\mathrm{S2}}. \tag{3.268}$$

If we don't care much about white noise amplification, we can derive other supercardioid beamformers that give higher values of the FBR than the one

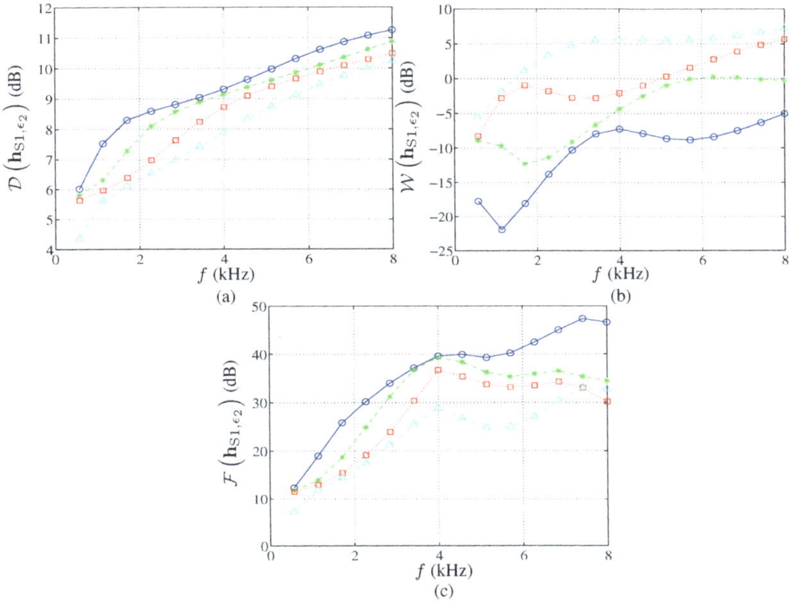

Fig. 3.34 Performance of the robust global supercardioid, $\mathbf{h}_{S1,\epsilon_2}$, as a function of frequency for $\delta = 5$ mm, $M_0 = 4$, and several values of ϵ_2: $\epsilon_2 = 10^{-5}$ (solid line with circles), $\epsilon_2 = 10^{-4}$ (dashed line with asterisks), $\epsilon_2 = 10^{-3}$ (dotted line with squares), and $\epsilon_2 = 10^{-2}$ (dash-dot line with triangles). (a) DF, (b) WNG, and (c) FBR.

with $\mathbf{h}_{S1}$ or $\mathbf{h}_{S2}$. For example, we can maximize separately the two FBRs, $\mathcal{F}_1(\mathbf{h}_1)$ and $\mathcal{F}_2(\mathbf{h}_2)$, of the subarrays. We get

$$\mathbf{h}_{1,S} = \frac{\mathbf{t}_1}{\mathbf{d}_{1,0}^H \mathbf{t}_1}, \tag{3.269}$$

$$\mathbf{h}_{2,S} = \frac{\mathbf{t}_2}{\mathbf{d}_{2,0}^H \mathbf{t}_2}, \tag{3.270}$$

where $\mathbf{t}_1$ and $\mathbf{t}_2$ are the eigenvectors corresponding to the maximum eigenvalues of the matrices $\mathbf{\Gamma}_{b,1}^{-1}\mathbf{\Gamma}_{f,1}$ and $\mathbf{\Gamma}_{b,2}^{-1}\mathbf{\Gamma}_{f,2}$, respectively. As a consequence, another global supercardioid is

$$\mathbf{h}_{S3} = \mathbf{h}_{1,S} \otimes \mathbf{h}_{2,S}. \tag{3.271}$$

Figure 3.35 displays the directivity patterns of the robust global supercardioid, $\mathbf{h}_{S3}$, for $f = 1$ kHz, $\delta = 5$ mm, and different numbers of sensors M. Figure 3.36 shows plots of the DFs, WNGs, and FBRs of the robust global supercardioid, $\mathbf{h}_{S3}$, as a function of frequency. We observe that the FBR and DF of $\mathbf{h}_{S3}$ are larger than those of $\mathbf{h}_{S1}$, but the WNG of $\mathbf{h}_{S3}$ is lower than that of $\mathbf{h}_{S1}$ (compare Figs 3.32 and 3.36). As we increase the number of sen-

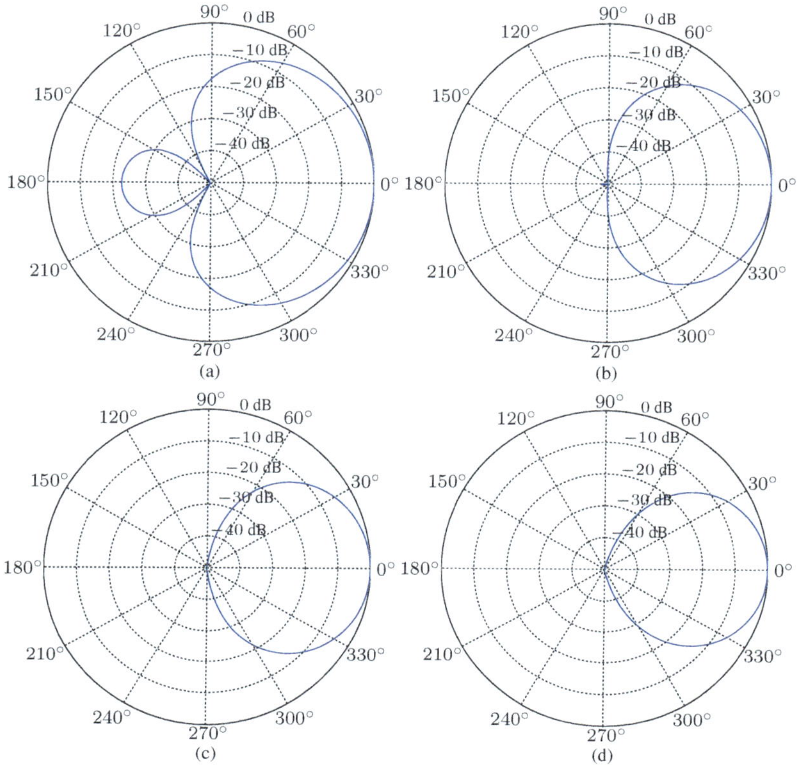

Fig. 3.35 Beampatterns of the robust global supercardioid, $\mathbf{h}_{\mathrm{S3}}$, for $f = 1$ kHz, $\delta = 5$ mm, and different numbers of sensors $M = M_0^2$: (a) $M_0 = 2$, (b) $M_0 = 3$, (c) $M_0 = 4$, and (d) $M_0 = 5$.

sors, the DF and FBR of the robust global supercardioid increase for the mid-range frequencies, but the WNG decreases.

Now, if we want to fully maximize the FBR in (3.206), we need to derive an iterative algorithm.

At iteration 0, we may take

$$\mathbf{h}_2^{(0)} = \mathbf{h}_{2,\mathrm{S}} = \frac{\mathbf{t}_2}{\mathbf{d}_{2,0}^H \mathbf{t}_2}. \quad (3.272)$$

Substituting $\mathbf{h}_2^{(0)}$ into (3.223) and (3.224), we get

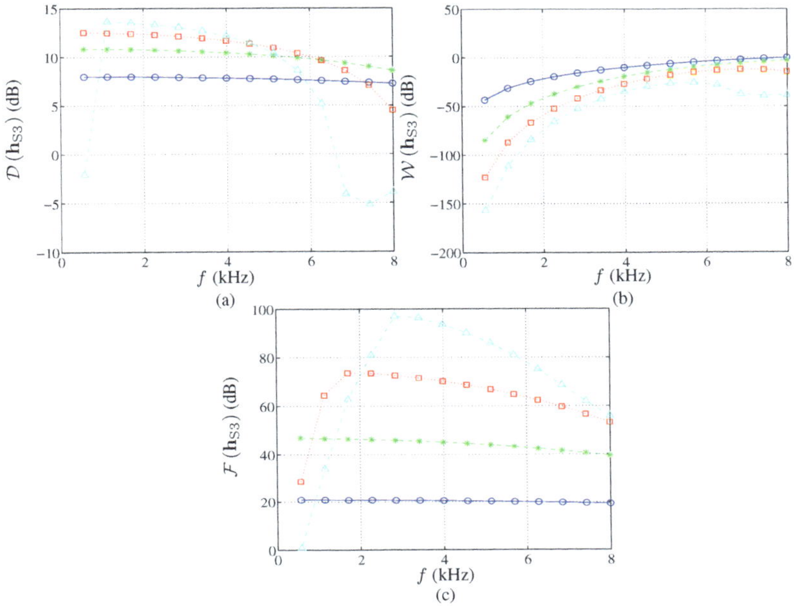

Fig. 3.36 Performance of the robust global supercardioid, $\mathbf{h}_{S3}$, as a function of frequency for $\delta = 5$ mm and different numbers of sensors $M = M_0^2$: $M_0 = 2$ (solid line with circles), $M_0 = 3$ (dashed line with asterisks), $M_0 = 4$ (dotted line with squares), and $M_0 = 5$ (dash-dot line with triangles). (a) DF, (b) WNG, and (c) FBR.

$$\mathbf{\Gamma}_{\mathrm{f},\mathbf{h}_2^{(0)}} = \left(\mathbf{I}_{M_0} \otimes \mathbf{h}_2^{(0)}\right)^H \mathbf{\Gamma}_{\mathrm{f}} \left(\mathbf{I}_{M_0} \otimes \mathbf{h}_2^{(0)}\right), \tag{3.273}$$

$$\mathbf{\Gamma}_{\mathrm{b},\mathbf{h}_2^{(0)}} = \left(\mathbf{I}_{M_0} \otimes \mathbf{h}_2^{(0)}\right)^H \mathbf{\Gamma}_{\mathrm{b}} \left(\mathbf{I}_{M_0} \otimes \mathbf{h}_2^{(0)}\right). \tag{3.274}$$

Now, plugging these expressions into the FBR in (3.222), we obtain at iteration 1:

$$\mathcal{F}\left(\mathbf{h}_1^{(1)}|\mathbf{h}_2^{(0)}\right) = \frac{\left(\mathbf{h}_1^{(1)}\right)^H \mathbf{\Gamma}_{\mathrm{f},\mathbf{h}_2^{(0)}} \mathbf{h}_1^{(1)}}{\left(\mathbf{h}_1^{(1)}\right)^H \mathbf{\Gamma}_{\mathrm{b},\mathbf{h}_2^{(0)}} \mathbf{h}_1^{(1)}}. \tag{3.275}$$

The maximization of $\mathcal{F}\left(\mathbf{h}_1^{(1)}|\mathbf{h}_2^{(0)}\right)$ with respect of $\mathbf{h}_1^{(1)}$ leads to

$$\mathbf{h}_1^{(1)} = \frac{\mathbf{t}_1^{(0)}}{\mathbf{d}_{1,0}^H \mathbf{t}_1^{(0)}}, \tag{3.276}$$

where $\mathbf{t}_1^{(0)}$ is the eigenvector corresponding to the maximum eigenvalue of the matrix $\mathbf{\Gamma}^{-1}_{\mathrm{b},\mathbf{h}_2^{(0)}} \mathbf{\Gamma}_{\mathrm{f},\mathbf{h}_2^{(0)}}$. Using $\mathbf{h}_1^{(1)}$ in (3.226) and (3.227), we get

$$\mathbf{\Gamma}_{\mathrm{f},\mathbf{h}_1^{(1)}} = \left(\mathbf{h}_1^{(1)} \otimes \mathbf{I}_{M_0}\right)^H \mathbf{\Gamma}_{\mathrm{f}} \left(\mathbf{h}_1^{(1)} \otimes \mathbf{I}_{M_0}\right), \tag{3.277}$$

$$\mathbf{\Gamma}_{\mathrm{b},\mathbf{h}_1^{(1)}} = \left(\mathbf{h}_1^{(1)} \otimes \mathbf{I}_{M_0}\right)^H \mathbf{\Gamma}_{\mathrm{b}} \left(\mathbf{h}_1^{(1)} \otimes \mathbf{I}_{M_0}\right). \tag{3.278}$$

As a result, the FBR in (3.225) is

$$\mathcal{F}\left(\mathbf{h}_2^{(1)}|\mathbf{h}_1^{(1)}\right) = \frac{\left(\mathbf{h}_2^{(1)}\right)^H \mathbf{\Gamma}_{\mathrm{f},\mathbf{h}_1^{(1)}} \mathbf{h}_2^{(1)}}{\left(\mathbf{h}_2^{(1)}\right)^H \mathbf{\Gamma}_{\mathrm{b},\mathbf{h}_1^{(1)}} \mathbf{h}_2^{(1)}}, \tag{3.279}$$

whose maximization with respect to $\mathbf{h}_2^{(1)}$ gives

$$\mathbf{h}_2^{(1)} = \frac{\mathbf{t}_2^{(1)}}{\mathbf{d}_{2,0}^H \mathbf{t}_2^{(1)}}, \tag{3.280}$$

where $\mathbf{t}_2^{(1)}$ is the eigenvector corresponding to the maximum eigenvalue of the matrix $\mathbf{\Gamma}^{-1}_{\mathrm{b},\mathbf{h}_1^{(1)}} \mathbf{\Gamma}_{\mathrm{f},\mathbf{h}_1^{(1)}}$.

Continuing to iterate up to iteration n, we easily get for the first filter:

$$\mathbf{h}_1^{(n)} = \frac{\mathbf{t}_1^{(n-1)}}{\mathbf{d}_{1,0}^H \mathbf{t}_1^{(n-1)}}, \tag{3.281}$$

where $\mathbf{t}_1^{(n-1)}$ is the eigenvector corresponding to the maximum eigenvalue of the matrix $\mathbf{\Gamma}^{-1}_{\mathrm{b},\mathbf{h}_2^{(n-1)}} \mathbf{\Gamma}_{\mathrm{f},\mathbf{h}_2^{(n-1)}}$, with

$$\mathbf{\Gamma}_{\mathrm{f},\mathbf{h}_2^{(n-1)}} = \left(\mathbf{I}_{M_0} \otimes \mathbf{h}_2^{(n-1)}\right)^H \mathbf{\Gamma}_{\mathrm{f}} \left(\mathbf{I}_{M_0} \otimes \mathbf{h}_2^{(n-1)}\right), \tag{3.282}$$

$$\mathbf{\Gamma}_{\mathrm{b},\mathbf{h}_2^{(n-1)}} = \left(\mathbf{I}_{M_0} \otimes \mathbf{h}_2^{(n-1)}\right)^H \mathbf{\Gamma}_{\mathrm{b}} \left(\mathbf{I}_{M_0} \otimes \mathbf{h}_2^{(n-1)}\right), \tag{3.283}$$

and for the second filter:

$$\mathbf{h}_2^{(n)} = \frac{\mathbf{t}_2^{(n)}}{\mathbf{d}_{2,0}^H \mathbf{t}_2^{(n)}}, \tag{3.284}$$

where $\mathbf{t}_2^{(n)}$ is the eigenvector corresponding to the maximum eigenvalue of the matrix $\mathbf{\Gamma}^{-1}_{\mathrm{b},\mathbf{h}_1^{(n)}} \mathbf{\Gamma}_{\mathrm{f},\mathbf{h}_1^{(n)}}$, with

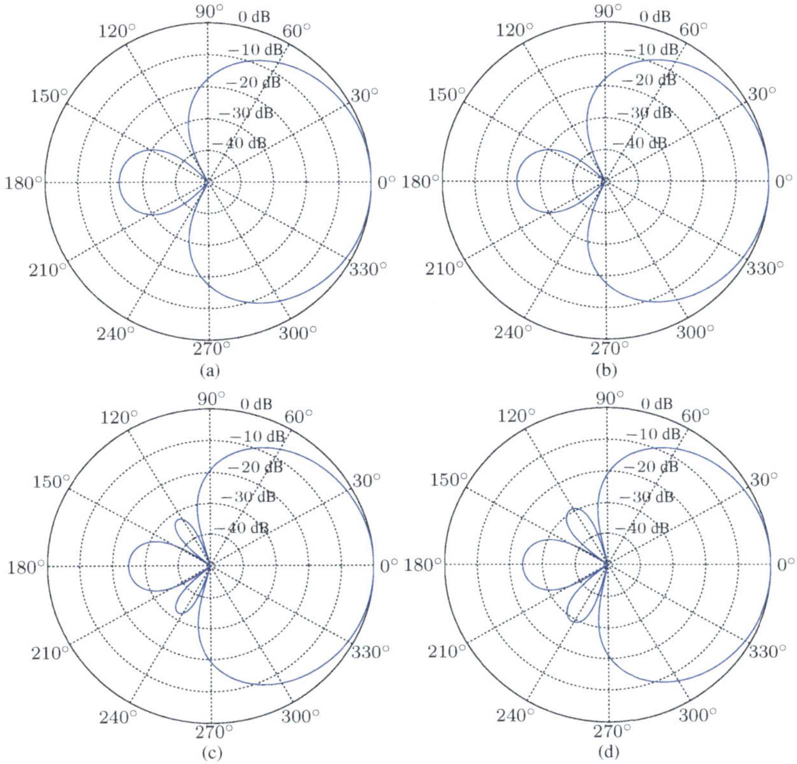

Fig. 3.37 Beampatterns of the supercardioid beamformer, $\mathbf{h}_{\mathrm{S}}^{(n)}$, for $M_0 = 2$, $f = 1$ kHz, and $\delta = 5$ mm, obtained at the iteration n: (a) $n = 0$, (b) $n = 1$, (c) $n = 2$, and (d) $n = 5$.

$$\mathbf{\Gamma}_{\mathrm{f},\mathbf{h}_1^{(n)}} = \left(\mathbf{h}_1^{(n)} \otimes \mathbf{I}_{M_0}\right)^H \mathbf{\Gamma}_{\mathrm{f}} \left(\mathbf{h}_1^{(n)} \otimes \mathbf{I}_{M_0}\right), \tag{3.285}$$

$$\mathbf{\Gamma}_{\mathrm{b},\mathbf{h}_1^{(n)}} = \left(\mathbf{h}_1^{(n)} \otimes \mathbf{I}_{M_0}\right)^H \mathbf{\Gamma}_{\mathrm{b}} \left(\mathbf{h}_1^{(n)} \otimes \mathbf{I}_{M_0}\right). \tag{3.286}$$

Finally, we deduce that the supercardioid beamformer is at iteration n:

$$\mathbf{h}_{\mathrm{S}}^{(n)} = \mathbf{h}_1^{(n)} \otimes \mathbf{h}_2^{(n)}. \tag{3.287}$$

Figure 3.37 displays the directivity patterns of the supercardioid beamformer, $\mathbf{h}_{\mathrm{S}}^{(n)}$, for $M_0 = 2$, $f = 1$ kHz, and $\delta = 5$ mm, obtained at the iteration n for several values of n. Figure 3.38 shows plots of the DFs, WNGs, and FBRs of the supercardioid beamformer, $\mathbf{h}_{\mathrm{S}}^{(n)}$, as a function of frequency. The iteration $n = 0$ corresponds to the robust global supercardioid, $\mathbf{h}_{\mathrm{S3}}$. We observe that the FBR and DF of the supercardioid beamformer, $\mathbf{h}_{\mathrm{S}}^{(n)}$, are larger than those of $\mathbf{h}_{\mathrm{S3}}$ and $\mathbf{h}_{\mathrm{S1}}$, but the WNGs of $\mathbf{h}_{\mathrm{S}}^{(n)}$ and $\mathbf{h}_{\mathrm{S3}}$ are lower

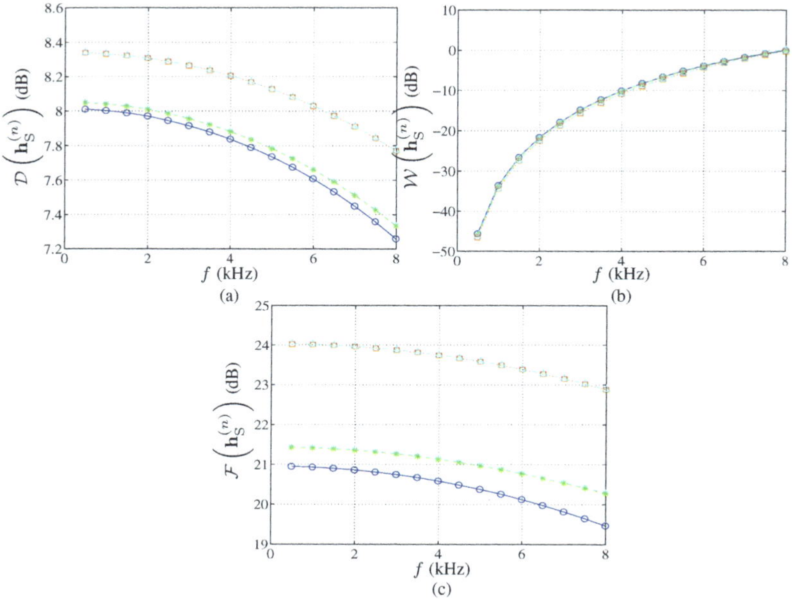

Fig. 3.38 Performance of the supercardioid beamformer, $\mathbf{h}_{\mathrm{S}}^{(n)}$, as a function of frequency for $M_0 = 2$, $\delta = 5$ mm, and several values of n: $n = 0$ (solid line with circles), $n = 1$ (dashed line with asterisks), $n = 3$ (dotted line with squares), and $n = 10$ (dash-dot line with triangles). (a) DF, (b) WNG, and (c) FBR.

than that of $\mathbf{h}_{\mathrm{S1}}$ (compare Figs 3.38, 3.36 and 3.32). Furthermore, the DF and FBR of the supercardioid beamformer, $\mathbf{h}_{\mathrm{S}}^{(n)}$, increase at each iteration, and roughly converge after three iterations, while the WNG remains almost the same at each iteration.

References

1. H. L. Van Trees, *Optimum Array Processing: Part IV of Detection, Estimation, and Modulation Theory*. New York, NY: John Wiley & Sons, Inc., 2002.
2. J. Benesty, I. Cohen, and J. Chen, *Fundamentals of Signal Enhancement and Array Signal Processing*. Singapore: Wiley–IEEE Press, 2018.
3. S. Haykin, *Adaptive Filter Theory*. Fourth Edition, Upper Saddle River, NJ: Prentice-Hall, 2002.
4. J. Benesty, C. Paleologu, and S. Ciochina, "On the identification of bilinear forms with the Wiener filter," *IEEE Signal Process. Lett.*, vol. 24, pp. 653–657, May 2017.
5. J. Capon, "High resolution frequency-wavenumber spectrum analysis," *Proc. IEEE*, vol. 57, pp. 1408–1418, Aug. 1969.
6. R. T. Lacoss, "Data adaptive spectral analysis methods," *Geophysics*, vol. 36, pp. 661–675, Aug. 1971.

7. A. Booker and C. Y. Ong, "Multiple constraint adaptive filtering," *Geophysics*, vol. 36, pp. 498–509, June 1971.
8. O. Frost, "An algorithm for linearly constrained adaptive array processing," *Proc. IEEE*, vol. 60, pp. 926–935, Jan. 1972.
9. G. W. Elko and J. Meyer, "Microphone arrays," in *Springer Handbook of Speech Processing*, J. Benesty, M. M. Sondhi, and Y. Huang, Eds., Berlin, Germany: Springer-Verlag, 2008, Chapter 50, pp. 1021–1041.
10. G. W. Elko, "Superdirectional microphone arrays," in *Acoustic Signal Processing for Telecommunication*, S. L. Gay and J. Benesty, Eds. Boston, MA: Kluwer Academic Publishers, 2000, Chapter 10, pp. 181–237.
11. J. Benesty and J. Chen, *Study and Design of Differential Microphone Arrays*. Berlin, Germany: Springer-Verlag, 2012.
12. J. Chen, J. Benesty, and C. Pan "On the design and implementation of linear differential microphone arrays," *J. Acoust. Soc. Am.*, vol. 136, pp. 3097–3113, Dec. 2014.
13. R. N. Marshall and W. R. Harry, "A new microphone providing uniform directivity over an extended frequency range," *J. Acoust. Soc. Am.*, vol. 12, pp. 481–497, 1941.

Chapter 4
Generalization with Uniform Linear Arrays

In this chapter, we show that other decompositions of the steering vector associated with ULAs are possible. Then, we define beamformers accordingly and derive all performance measures useful for fixed beamforming. A particular decomposition has caught our attention, for which we explain in details how it may fit well in differential beamforming. Obviously, we can follow all these steps to deduce beamformers for other decompositions.

4.1 Signal Model and Problem Formulation

As in Chapter 2, we are still considering a ULA of size $(M-1)\delta$. We recall that the steering vector is [1], [2]

$$\mathbf{d}_\theta = \begin{bmatrix} 1 & e^{-\jmath\varpi(\theta)} & e^{-\jmath 2\varpi(\theta)} & \cdots & e^{-\jmath(M-1)\varpi(\theta)} \end{bmatrix}^T, \tag{4.1}$$

where

$$\varpi(\theta) = \frac{\omega\delta\cos\theta}{c}. \tag{4.2}$$

In the two previous chapters, we focused on the very particular case where the number of omnidirectional microphones forming the array is a square number, i.e., $M = M_0^2$ with $M_0 \geq 2$. Then, we found that the steering vector can be decomposed as

$$\mathbf{d}_\theta = \mathbf{d}_{1,\theta} \otimes \mathbf{d}_{2,\theta}, \tag{4.3}$$

where

J. Benesty et al., *Array Processing*, Springer Topics in Signal Processing 18, https://doi.org/10.1007/978-3-030-15600-8_4

$$\mathbf{d}_{1,\theta} = \begin{bmatrix} 1 & e^{-\jmath M_0 \varpi(\theta)} & e^{-\jmath 2 M_0 \varpi(\theta)} & \cdots & e^{-\jmath M_0 (M_0 - 1)\varpi(\theta)} \end{bmatrix}^T, \tag{4.4}$$

$$\mathbf{d}_{2,\theta} = \begin{bmatrix} 1 & e^{-\jmath \varpi(\theta)} & e^{-\jmath 2 \varpi(\theta)} & \cdots & e^{-\jmath (M_0 - 1)\varpi(\theta)} \end{bmatrix}^T. \tag{4.5}$$

Clearly, other decompositions or generalizations are possible. An obvious one that comes immediately in mind is when $M = M_1 M_2$. In this case, we have

$$\mathbf{d}_\theta = \mathbf{d}_{1,\theta} \otimes \mathbf{d}_{2,\theta}, \tag{4.6}$$

where

$$\mathbf{d}_{1,\theta} = \begin{bmatrix} 1 & e^{-\jmath M_2 \varpi(\theta)} & e^{-\jmath 2 M_2 \varpi(\theta)} & \cdots & e^{-\jmath M_2 (M_1 - 1)\varpi(\theta)} \end{bmatrix}^T \tag{4.7}$$

is the steering vector (of length M_1) corresponding to a ULA of M_1 sensors with an interelement spacing equal to $M_2\delta$ and

$$\mathbf{d}_{2,\theta} = \begin{bmatrix} 1 & e^{-\jmath \varpi(\theta)} & e^{-\jmath 2 \varpi(\theta)} & \cdots & e^{-\jmath (M_2 - 1)\varpi(\theta)} \end{bmatrix}^T \tag{4.8}$$

is the steering vector (of length M_2) corresponding to a ULA of M_2 sensors with an interelement spacing equal to δ (some examples are shown in Fig. 1.2).

A much more general decomposition is possible when the number of sensors can be factorized as

$$M = M_1 \times M_2 \times \cdots \times M_P. \tag{4.9}$$

Then, it is not hard to see that we have

$$\mathbf{d}_\theta = \mathbf{d}_{1,\theta} \otimes \mathbf{d}_{2,\theta} \otimes \cdots \otimes \mathbf{d}_{P,\theta}, \tag{4.10}$$

where

$$\mathbf{d}_{p,\theta} = \begin{bmatrix} 1 & e^{-\jmath \nu_p \varpi(\theta)} & e^{-\jmath 2 \nu_p \varpi(\theta)} & \cdots & e^{-\jmath \nu_p (M_p - 1)\varpi(\theta)} \end{bmatrix}^T, \tag{4.11}$$

for $p = 1, 2, \ldots, P$, is the steering vector (of length M_p) corresponding to the pth ULA of M_p sensors with an interelement spacing equal to $\nu_p\delta$, and

$$\nu_p = M_{p+1} \times M_{p+2} \times \cdots \times M_P, \tag{4.12}$$

with $M_{P+1} = 1$, so that $\nu_P = 1$. The decomposition in (4.10) is not new; similar ones were proposed in the literature and in different contexts. See for example [3].

In this chapter, we will study beamforming in the particular scenario where $M_1 = M_2 = \cdots = M_P = 2$, so that $M = 2^P$. As a result, the steering vector can be decomposed as in (4.10), where

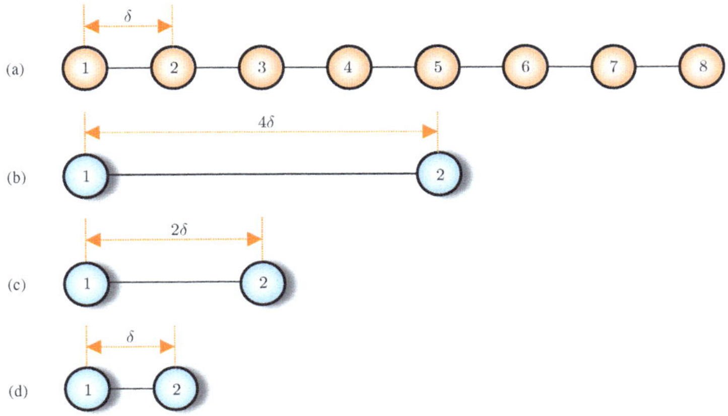

Fig. 4.1 Kronecker product decomposition of a ULA of 8 sensors into three virtual ULAs of 2 sensors each: (a) global array with an interelement spacing equal to δ; (b) 1st ULA of 2 sensors with an interelement spacing equal to 4δ; (c) 2nd ULA of 2 sensors with an interelement spacing equal to 2δ; and (d) 3rd ULA of 2 sensors with an interelement spacing equal to δ.

$$\mathbf{d}_{p,\theta} = \begin{bmatrix} 1 \\ e^{-\jmath 2^{P-p}\varpi(\theta)} \end{bmatrix}, \tag{4.13}$$

for $p = 1, 2, \ldots, P$, is the steering vector corresponding to the pth ULA of two sensors with an interelement spacing equal to $2^{P-p}\delta$. Figure 4.1 shows an example of Kronecker product decomposition of a ULA of 8 sensors into three virtual ULAs of 2 sensors each.

We assume that δ is small and the angle of the desired signal is $\theta_{\mathrm{d}} = 0$, so that the focus is on differential beamforming only. The signal model is then [4], [5]

$$\mathbf{y} = \mathbf{d}_0 X + \mathbf{v}, \tag{4.14}$$

where $\mathbf{d}_0 = \mathbf{d}_{1,0} \otimes \mathbf{d}_{2,0} \otimes \cdots \otimes \mathbf{d}_{P,0}$. We deduce that the covariance matrix of $\mathbf{y}$ is

$$\mathbf{\Phi}_{\mathbf{y}} = \phi_X \mathbf{d}_0 \mathbf{d}_0^H + \phi_{V_1} \mathbf{\Gamma}_{\mathbf{v}}. \tag{4.15}$$

Our objective in this chapter is to design, when $M = 2^P$, the most important differential beamforming beampatterns (of, roughly, order P) thanks to the Kronecker product filtering approach.

4.2 Beamforming with Kronecker Product Filters

As we have already explained in Chapter 2, we consider complex-valued filters that follow the decomposition of the global steering vector, $\mathbf{d}_\theta$. Here, they are of length $M = 2^P$ and can be factorized as

$$\mathbf{h} = \mathbf{h}_1 \otimes \mathbf{h}_2 \otimes \cdots \otimes \mathbf{h}_P, \tag{4.16}$$

where $\mathbf{h}_p$, $p = 1, 2, \ldots, P$ are P complex-valued linear filters of length 2. Then, beamforming is performed by applying $\mathbf{h}$ to $\mathbf{y}$, i.e.,

$$\begin{aligned} Z &= \mathbf{h}^H \mathbf{y} \\ &= \mathbf{h}^H \mathbf{d}_0 X + \mathbf{h}^H \mathbf{v} \\ &= X_{\mathrm{fd}} + V_{\mathrm{rn}}, \end{aligned} \tag{4.17}$$

where Z is the estimate of the desired signal, X,

$$X_{\mathrm{fd}} = X \prod_{p=1}^{P} \mathbf{h}_p^H \mathbf{d}_{p,0} \tag{4.18}$$

is the filtered desired signal, and

$$V_{\mathrm{rn}} = \mathbf{h}^H \mathbf{v} \tag{4.19}$$

is the residual noise. We deduce that the variance of Z is

$$\phi_Z = \phi_X \prod_{p=1}^{P} \left| \mathbf{h}_p^H \mathbf{d}_{p,0} \right|^2 + \phi_{V_1} \mathbf{h}^H \mathbf{\Gamma}_{\mathbf{v}} \mathbf{h}. \tag{4.20}$$

With Kronecker product beamforming, we only need to estimate $2P$ parameters instead of 2^P with conventional beamforming.

Since in the design or optimization process of beampatterns, it is much more convenient to work with the individual filters $\mathbf{h}_p$, $p = 1, 2, \ldots, P$ instead of the global beamformer $\mathbf{h}$, we only consider the individual distortionless constraints in the direction of the desired signal, $\theta = 0$. They are

$$\mathbf{h}_p^H \mathbf{d}_{p,0} = 1, \ p = 1, 2, \ldots, P. \tag{4.21}$$

Obviously, when the previous constraints are fulfilled, then $\mathbf{h}^H \mathbf{d}_0 = 1$.

4.3 Performance Measures

In this section, we recall the most useful performance measures in this context for fixed beamforming.

The beampattern of $\mathbf{h}$ is defined as

$$\begin{aligned}\mathcal{B}_\theta(\mathbf{h}) &= \mathbf{d}_\theta^H \mathbf{h} \\ &= \prod_{p=1}^{P} \mathbf{d}_{p,\theta}^H \mathbf{h}_p \\ &= \prod_{p=1}^{P} \mathcal{B}_{p,\theta}(\mathbf{h}_p),\end{aligned} \tag{4.22}$$

where

$$\begin{aligned}\mathcal{B}_{p,\theta}(\mathbf{h}_p) &= \mathbf{d}_{p,\theta}^H \mathbf{h}_p \\ &= \sum_{m=1}^{2} H_{p,m} e^{\jmath(m-1)2^{P-p}\varpi(\theta)}\end{aligned} \tag{4.23}$$

is the beampattern of the pth ULA, with $H_{p,m}$, $m = 1, 2$ being the coefficients of $\mathbf{h}_p$. Let $\mathcal{Z}_p = e^{\jmath 2^{P-p}\varpi(\theta)}$, we can express the global beampattern as a polynomial in P variables, which is the product of P polynomials (of degree 1) in one variable each, i.e.,

$$\begin{aligned}\mathcal{B}(\mathcal{Z}_1, \mathcal{Z}_2, \ldots, Z_P) &= \prod_{p=1}^{P} \mathcal{B}_p(\mathcal{Z}_p) \\ &= \prod_{p=1}^{P} (H_{p,1} + H_{p,2} Z_p).\end{aligned} \tag{4.24}$$

From this perspective, we can see that this beampattern has at most P distinct nulls (between 0 and π).

It is easy to see that the gain in SNR is

$$\mathcal{G}(\mathbf{h}) = \frac{\left|\mathbf{h}^H \mathbf{d}_0\right|^2}{\mathbf{h}^H \mathbf{\Gamma}_\mathbf{v} \mathbf{h}}. \tag{4.25}$$

If the nature of the noise is such that its pseudo-coherence matrix can be factorized as $\mathbf{\Gamma}_\mathbf{v} = \mathbf{\Gamma}_{\mathbf{v},1} \otimes \mathbf{\Gamma}_{\mathbf{v},2} \otimes \cdots \otimes \mathbf{\Gamma}_{\mathbf{v},P}$, where $\mathbf{\Gamma}_{\mathbf{v},p}$ is the pseudo-coherence matrix of the noise at the pth ULA, then the gain in SNR simplifies to

$$\mathcal{G}(\mathbf{h}) = \frac{\prod_{p=1}^{P} \left|\mathbf{h}_p^H \mathbf{d}_{p,0}\right|^2}{\prod_{p=1}^{P} \mathbf{h}_p^H \boldsymbol{\Gamma}_{\mathbf{v},p} \mathbf{h}_p}$$
$$= \prod_{p=1}^{P} \mathcal{G}_p(\mathbf{h}_p), \tag{4.26}$$

where

$$\mathcal{G}_p(\mathbf{h}_p) = \frac{\left|\mathbf{h}_p^H \mathbf{d}_{p,0}\right|^2}{\mathbf{h}_p^H \boldsymbol{\Gamma}_{\mathbf{v},p} \mathbf{h}_p}. \tag{4.27}$$

One important particular case of the gain in SNR is the WNG, i.e.,

$$\mathcal{W}(\mathbf{h}) = \frac{\left|\mathbf{h}^H \mathbf{d}_0\right|^2}{\mathbf{h}^H \mathbf{h}} \tag{4.28}$$
$$= \frac{\prod_{p=1}^{P} \left|\mathbf{h}_p^H \mathbf{d}_{p,0}\right|^2}{\prod_{p=1}^{P} \mathbf{h}_p^H \mathbf{h}_p}$$
$$= \prod_{p=1}^{P} \mathcal{W}_p(\mathbf{h}_p),$$

where

$$\mathcal{W}_p(\mathbf{h}_p) = \frac{\left|\mathbf{h}_p^H \mathbf{d}_{p,0}\right|^2}{\mathbf{h}_p^H \mathbf{h}_p}. \tag{4.29}$$

Obviously, the WNG of the global ULA is simply the product of the WNGs of the P ULAs. It is clear that

$$\mathcal{W}(\mathbf{h}) \leq 2^P, \ \forall \mathbf{h}. \tag{4.30}$$

Another important particular case of the gain in SNR is the DF, i.e.,

$$\mathcal{D}(\mathbf{h}) = \frac{\left|\mathcal{B}_0(\mathbf{h})\right|^2}{\frac{1}{2}\int_0^{\pi} \left|\mathcal{B}_\theta(\mathbf{h})\right|^2 \sin\theta d\theta} \tag{4.31}$$
$$= \frac{\prod_{p=1}^{P} \left|\mathcal{B}_{p,0}(\mathbf{h}_p)\right|^2}{\frac{1}{2}\int_0^{\pi} \prod_{p=1}^{P} \left|\mathcal{B}_{p,\theta}(\mathbf{h}_p)\right|^2 \sin\theta d\theta}$$
$$= \frac{\left|\mathbf{h}^H \mathbf{d}_0\right|^2}{\mathbf{h}^H \boldsymbol{\Gamma} \mathbf{h}},$$

where

$$\mathbf{\Gamma} = \frac{1}{2} \int_0^{\pi} \mathbf{d}_\theta \mathbf{d}_\theta^H \sin\theta d\theta, \tag{4.32}$$

whose elements are given by

$$[\mathbf{\Gamma}(\omega)]_{ij} = \text{sinc}\left[\omega(j-i)\delta/c\right]. \tag{4.33}$$

It is clear that

$$\mathcal{D}(\mathbf{h}) \leq \mathbf{d}_0^H \mathbf{\Gamma}^{-1} \mathbf{d}_0, \ \forall \mathbf{h}. \tag{4.34}$$

As we already know, the DF of the global beamformer cannot be factorized, i.e.,

$$\mathcal{D}(\mathbf{h}) \neq \prod_{p=1}^{P} \mathcal{D}_p(\mathbf{h}_p), \tag{4.35}$$

where

$$\begin{aligned} \mathcal{D}_p(\mathbf{h}_p) &= \frac{\left|\mathcal{B}_{p,0}(\mathbf{h}_p)\right|^2}{\frac{1}{2}\int_0^{\pi} \left|\mathcal{B}_{p,\theta}(\mathbf{h}_p)\right|^2 \sin\theta d\theta} \\ &= \frac{\left|\mathbf{h}_p^H \mathbf{d}_{p,0}\right|^2}{\mathbf{h}_p^H \mathbf{\Gamma}_p \mathbf{h}_p}, \end{aligned} \tag{4.36}$$

with

$$\mathbf{\Gamma}_p = \frac{1}{2} \int_0^{\pi} \mathbf{d}_{p,\theta} \mathbf{d}_{p,\theta}^H \sin\theta d\theta, \tag{4.37}$$

whose elements are given by

$$[\mathbf{\Gamma}_p(\omega)]_{ij} = \text{sinc}\left[\omega(j-i)2^{P-p}\delta/c\right]. \tag{4.38}$$

The last measure of interest in this section is the FBR defined as

$$\mathcal{F}(\mathbf{h}) = \frac{\int_0^{\pi/2} |\mathcal{B}_\theta(\mathbf{h})|^2 \sin\theta d\theta}{\int_{\pi/2}^{\pi} |\mathcal{B}_\theta(\mathbf{h})|^2 \sin\theta d\theta} \tag{4.39}$$
$$= \frac{\int_0^{\pi/2} \prod_{p=1}^{P} |\mathcal{B}_{p,\theta}(\mathbf{h}_p)|^2 \sin\theta d\theta}{\int_{\pi/2}^{\pi} \prod_{p=1}^{P} |\mathcal{B}_{p,\theta}(\mathbf{h}_p)|^2 \sin\theta d\theta}$$
$$= \frac{\mathbf{h}^H \boldsymbol{\Gamma}_\mathrm{f} \mathbf{h}}{\mathbf{h}^H \boldsymbol{\Gamma}_\mathrm{b} \mathbf{h}},$$

where

$$\boldsymbol{\Gamma}_\mathrm{f} = \int_0^{\pi/2} \mathbf{d}_\theta \mathbf{d}_\theta^H \sin\theta d\theta, \tag{4.40}$$

$$\boldsymbol{\Gamma}_\mathrm{b} = \int_{\pi/2}^{\pi} \mathbf{d}_\theta \mathbf{d}_\theta^H \sin\theta d\theta, \tag{4.41}$$

whose elements are given by

$$[\boldsymbol{\Gamma}_\mathrm{f}(\omega)]_{ij} = \frac{e^{\jmath\omega(j-i)\delta/c} - 1}{\jmath\omega(j-i)\delta/c}, \tag{4.42}$$

$$[\boldsymbol{\Gamma}_\mathrm{b}(\omega)]_{ij} = \frac{1 - e^{-\jmath\omega(j-i)\delta/c}}{\jmath\omega(j-i)\delta/c}. \tag{4.43}$$

Same as the DF, the FBR cannot be factorized, i.e.,

$$\mathcal{F}(\mathbf{h}) \neq \prod_{p=1}^{P} \mathcal{F}_p(\mathbf{h}_p), \tag{4.44}$$

where

$$\mathcal{F}_p(\mathbf{h}_p) = \frac{\int_0^{\pi/2} |\mathcal{B}_{p,\theta}(\mathbf{h}_p)|^2 \sin\theta d\theta}{\int_{\pi/2}^{\pi} |\mathcal{B}_{p,\theta}(\mathbf{h}_p)|^2 \sin\theta d\theta} \tag{4.45}$$
$$= \frac{\mathbf{h}_p^H \boldsymbol{\Gamma}_{\mathrm{f},p} \mathbf{h}_p}{\mathbf{h}_p^H \boldsymbol{\Gamma}_{\mathrm{b},p} \mathbf{h}_p},$$

with

$$\mathbf{\Gamma}_{\mathrm{f},p} = \int_0^{\pi/2} \mathbf{d}_{p,\theta}\mathbf{d}_{p,\theta}^H \sin\theta d\theta, \tag{4.46}$$

$$\mathbf{\Gamma}_{\mathrm{b},p} = \int_{\pi/2}^{\pi} \mathbf{d}_{p,\theta}\mathbf{d}_{p,\theta}^H \sin\theta d\theta, \tag{4.47}$$

whose elements are given by

$$[\mathbf{\Gamma}_{\mathrm{f},p}(\omega)]_{ij} = \frac{e^{\jmath\omega(j-i)2^{P-p}\delta/c} - 1}{\jmath\omega(j-i)2^{P-p}\delta/c}, \tag{4.48}$$

$$[\mathbf{\Gamma}_{\mathrm{b},p}(\omega)]_{ij} = \frac{1 - e^{-\jmath\omega(j-i)2^{P-p}\delta/c}}{\jmath\omega(j-i)2^{P-p}\delta/c}. \tag{4.49}$$

4.4 Differential Beamformers

In this section, we show how to design important differential beamformers, in a very elegant way, thanks to the above described decompositions of the steering vector and global filter.

4.4.1 Principle

We know from our past work that we can design, in a very simple and general fashion, all kinds of differential beamformers with null constraints [6], [7]. The direction of the null determines the shape of the beampattern. Here, since we have P filters, $\mathbf{h}_p$, $p = 1, 2, \ldots, P$, of length 2, each one of them can only handle one null. To simplify, in one of the proposed designs, we assume that all these filters will generate the same null in the direction θ_0 $(0 < \theta_0 \leq \pi)$; as a result, the global filter, $\mathbf{h}$, will generate a unique null of multiplicity P (in the direction θ_0) and the resulting differential beamformer will be of, roughly, order P. We see that for each filter, we have exactly two constraints to fulfill. The first constraint is the distortionless response (a one at the angle $\theta = 0$) and the second constraint is a null in the interval $0 < \theta \leq \pi$. Thus, these two constraints can be written as

$$\mathbf{d}_{p,0}^H \mathbf{h}_p = 1, \tag{4.50}$$

$$\mathbf{d}_{p,\theta_0}^H \mathbf{h}_p = 0. \tag{4.51}$$

We can express the previous equations as

$$\begin{bmatrix} \mathbf{d}_{p,0}^H \\ \mathbf{d}_{p,\theta_0}^H \end{bmatrix} \mathbf{h}_p = \begin{bmatrix} 1 & e^{\jmath 2^{P-p}\varpi(0)} \\ 1 & e^{\jmath 2^{P-p}\varpi(\theta_0)} \end{bmatrix} \mathbf{h}_p = \begin{bmatrix} 1 \\ 0 \end{bmatrix}, \tag{4.52}$$

which is a very simple linear system of equations to solve.

4.4.2 Dipole

The dipole has a one at the angle 0 and a null at the angle $\pi/2$. Hence, the linear system of two equations (4.52) becomes

$$\begin{bmatrix} 1 & e^{\jmath 2^{P-p}\varpi(0)} \\ 1 & 1 \end{bmatrix} \mathbf{h}_p = \begin{bmatrix} 1 \\ 0 \end{bmatrix}, \tag{4.53}$$

for which the solution is

$$\mathbf{h}_p = \frac{1}{1 - e^{\jmath 2^{P-p}\varpi(0)}} \begin{bmatrix} 1 \\ -1 \end{bmatrix}. \tag{4.54}$$

We deduce that the dipole of, roughly, order P is

$$\mathbf{h}_{\mathrm{D}} = \frac{1}{\prod_{p=1}^{P} \left[1 - e^{\jmath 2^{P-p}\varpi(0)}\right]} \begin{bmatrix} 1 \\ -1 \end{bmatrix} \otimes \begin{bmatrix} 1 \\ -1 \end{bmatrix} \otimes \cdots \otimes \begin{bmatrix} 1 \\ -1 \end{bmatrix}. \tag{4.55}$$

Thanks to (4.54), it is easy to see that the WNG and the power beampattern of $\mathbf{h}_{\mathrm{D}}$ are, respectively,

$$\begin{aligned} \mathcal{W}(\mathbf{h}_{\mathrm{D}}) &= \frac{1}{2^P} \prod_{p=1}^{P} \left|1 - e^{\jmath 2^{P-p}\varpi(0)}\right|^2 \\ &= \prod_{p=1}^{P} \left\{1 - \cos\left[2^{P-p}\varpi(0)\right]\right\} \end{aligned} \tag{4.56}$$

and

$$\begin{aligned} \left|\mathcal{B}(\mathbf{h}_{\mathrm{D}})\right|^2 &= \frac{\prod_{p=1}^{P} \left|1 - e^{\jmath 2^{P-p}\varpi(\theta)}\right|^2}{\prod_{p=1}^{P} \left|1 - e^{\jmath 2^{P-p}\varpi(0)}\right|^2} \\ &= \frac{\prod_{p=1}^{P} \left\{1 - \cos\left[2^{P-p}\varpi(\theta)\right]\right\}}{\prod_{p=1}^{P} \left\{1 - \cos\left[2^{P-p}\varpi(0)\right]\right\}}. \end{aligned} \tag{4.57}$$

Finally, the DF is

$$\mathcal{D}(\mathbf{h}_{\mathrm{D}}) = \frac{1}{\mathbf{h}_{\mathrm{D}}^H \mathbf{\Gamma} \mathbf{h}_{\mathrm{D}}}. \tag{4.58}$$

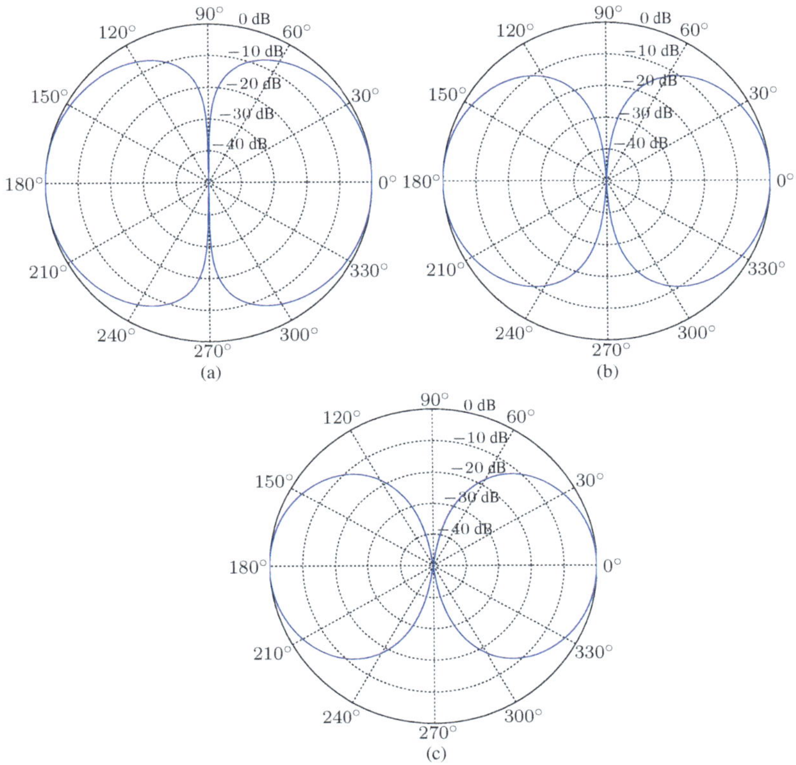

Fig. 4.2 Beampatterns of the dipole of order P for $f = 0.5$ kHz and $\delta = 1$ cm: (a) $P = 1$, (b) $P = 2$, and (c) $P = 3$.

Figure 4.2 displays the directivity patterns of the first-, second-, and third-order dipoles for $f = 0.5$ kHz and $\delta = 1$ cm. Figure 4.3 shows plots of the DFs and WNGs of the dipoles as a function of frequency. We observe that at low frequencies the DF increases as we increase the order of the dipole, but the WNG decreases. Of course, to get better compromises and performance, it is recommended to use different orders at different frequencies.

4.4.3 Cardioid

In the cardioid, there is a one at the angle 0 and a null at the angle π. Therefore, our linear system of two equations is

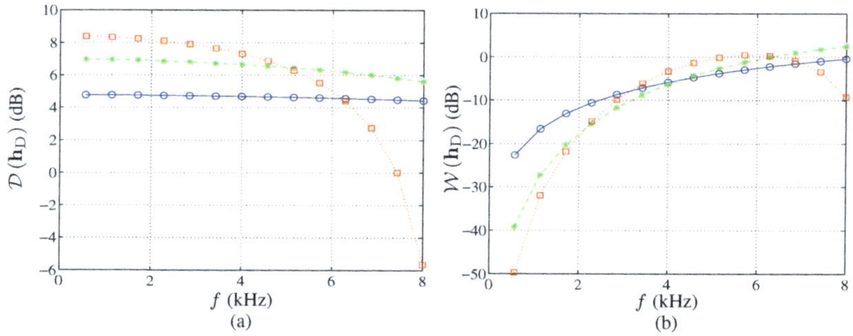

Fig. 4.3 Performance of the dipole of order P as a function of frequency for $\delta = 1$ cm and several values of P: $P = 1$ (solid line with circles), $P = 2$ (dashed line with asterisks), and $P = 3$ (dotted line with squares). (a) DF and (b) WNG.

$$\begin{bmatrix} 1 & e^{\jmath 2^{P-p}\varpi(0)} \\ 1 & e^{\jmath 2^{P-p}\varpi(\pi)} \end{bmatrix} \mathbf{h}_p = \begin{bmatrix} 1 & e^{\jmath 2^{P-p}\varpi(0)} \\ 1 & e^{-\jmath 2^{P-p}\varpi(0)} \end{bmatrix} \mathbf{h}_p = \begin{bmatrix} 1 \\ 0 \end{bmatrix}. \tag{4.59}$$

We easily find that the solution is

$$\begin{aligned} \mathbf{h}_p &= \frac{1}{e^{-\jmath 2^{P-p}\varpi(0)} - e^{\jmath 2^{P-p}\varpi(0)}} \begin{bmatrix} e^{-\jmath 2^{P-p}\varpi(0)} \\ -1 \end{bmatrix} \\ &= \frac{1}{1 - e^{\jmath 2^{P-p+1}\varpi(0)}} \begin{bmatrix} 1 \\ -e^{\jmath 2^{P-p}\varpi(0)} \end{bmatrix}. \end{aligned} \tag{4.60}$$

As a consequence, the cardioid of, roughly, order P is

$$\begin{aligned} \mathbf{h}_C = & \frac{1}{\prod_{p=1}^{P} \left[1 - e^{\jmath 2^{P-p+1}\varpi(0)}\right]} \times \\ & \begin{bmatrix} 1 \\ -e^{\jmath 2^{P-p}\varpi(0)} \end{bmatrix} \otimes \begin{bmatrix} 1 \\ -e^{\jmath 2^{P-p}\varpi(0)} \end{bmatrix} \otimes \cdots \otimes \begin{bmatrix} 1 \\ -e^{\jmath 2^{P-p}\varpi(0)} \end{bmatrix}. \end{aligned} \tag{4.61}$$

Thanks to (4.60), we can express the WNG and the power beampattern of $\mathbf{h}_C$ as, respectively,

$$\begin{aligned} \mathcal{W}(\mathbf{h}_C) &= \frac{1}{2^P} \prod_{p=1}^{P} \left| 1 - e^{\jmath 2^{P-p+1}\varpi(0)} \right|^2 \\ &= \prod_{p=1}^{P} \left\{ 1 - \cos\left[2 \times 2^{P-p}\varpi(0)\right] \right\} \end{aligned} \tag{4.62}$$

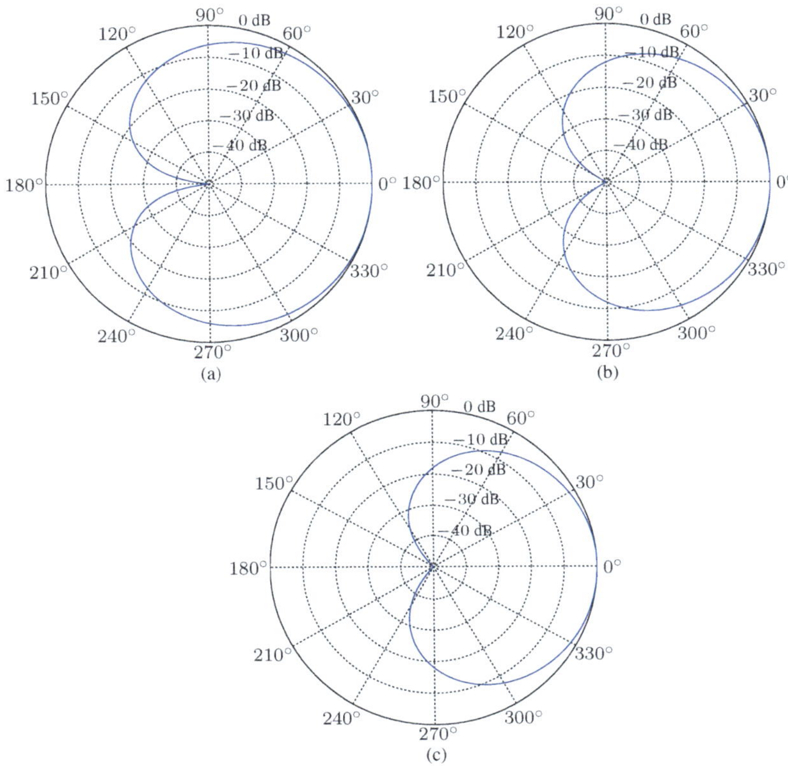

Fig. 4.4 Beampatterns of the cardioid of order P for $f = 0.5$ kHz and $\delta = 5$ mm: (a) $P = 1$, (b) $P = 2$, and (c) $P = 3$.

and

$$|\mathcal{B}(\mathbf{h}_{\mathrm{C}})|^2 = \frac{\prod_{p=1}^{P} \left|1 - e^{\jmath 2^{P-p+1}\varpi(\theta)}\right|^2}{\prod_{p=1}^{P} \left|1 - e^{\jmath 2^{P-p+1}\varpi(0)}\right|^2} = \frac{\prod_{p=1}^{P} \left\{1 - \cos\left[2 \times 2^{P-p}\varpi(\theta)\right]\right\}}{\prod_{p=1}^{P} \left\{1 - \cos\left[2 \times 2^{P-p}\varpi(0)\right]\right\}}. \tag{4.63}$$

Finally, the DF is

$$\mathcal{D}(\mathbf{h}_{\mathrm{C}}) = \frac{1}{\mathbf{h}_{\mathrm{C}}^H \mathbf{\Gamma} \mathbf{h}_{\mathrm{C}}}. \tag{4.64}$$

Figure 4.4 displays the directivity patterns of the first-, second-, and third-order cardioid for $f = 0.5$ kHz and $\delta = 5$ mm. Figure 4.5 shows plots of the

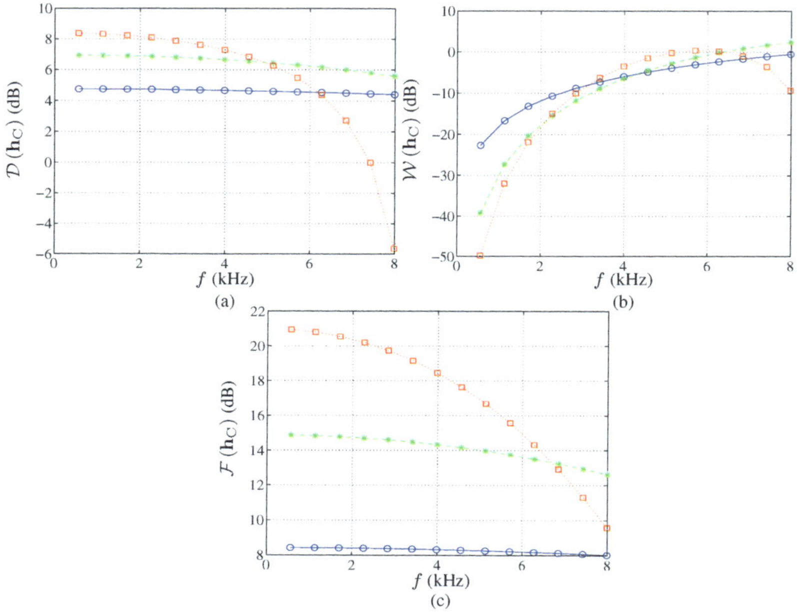

Fig. 4.5 Performance of the cardioid of order P as a function of frequency for $\delta = 5$ mm and several values of P: $P = 1$ (solid line with circles), $P = 2$ (dashed line with asterisks), and $P = 3$ (dotted line with squares). (a) DF, (b) WNG, and (c) FBR.

DFs, WNGs, and FBRs of the cardioids as a function of frequency. We observe that at low frequencies the DF and FBR increase as we increase the order of the cardioid, but the WNG decreases. If, for example, we want a good behavior of the DF, we can combine the second- and third-order cardioid, where the third order is used up to 4 kHz and the second order is used for above 4 kHz.

4.4.4 Hypercardioid

The first-order hypercardioid has a one at the angle 0 and a null at the angle $2\pi/3$. By simple extension, we can derive a kind of hypercardioid of, roughly, order P, as we did for the dipole or cardioid. Therefore, our linear system of two equations is

$$\begin{bmatrix} 1 & e^{\jmath 2^{P-p}\varpi(0)} \\ 1 & e^{\jmath 2^{P-p}\varpi(2\pi/3)} \end{bmatrix} \mathbf{h}_p = \begin{bmatrix} 1 & e^{\jmath 2^{P-p}\varpi(0)} \\ 1 & e^{-\jmath 2^{P-p-1}\varpi(0)} \end{bmatrix} \mathbf{h}_p = \begin{bmatrix} 1 \\ 0 \end{bmatrix}, \tag{4.65}$$

for which the solution is

$$\mathbf{h}_p = \frac{1}{e^{-\jmath 2^{P-p-1}\varpi(0)} - e^{\jmath 2^{P-p}\varpi(0)}} \begin{bmatrix} e^{-\jmath 2^{P-p-1}\varpi(0)} \\ -1 \end{bmatrix}$$
$$= \frac{1}{1 - e^{\jmath 3 \times 2^{P-p-1}\varpi(0)}} \begin{bmatrix} 1 \\ -e^{\jmath 2^{P-p-1}\varpi(0)} \end{bmatrix}. \quad (4.66)$$

Therefore, the first kind of hypercardioid of, roughly, order P that we propose here is

$$\mathbf{h}_{\mathrm{H1}} = \frac{1}{\prod_{p=1}^{P}\left[1 - e^{\jmath 3 \times 2^{P-p-1}\varpi(0)}\right]} \times$$
$$\begin{bmatrix} 1 \\ -e^{\jmath 2^{P-p-1}\varpi(0)} \end{bmatrix} \otimes \begin{bmatrix} 1 \\ -e^{\jmath 2^{P-p-1}\varpi(0)} \end{bmatrix} \otimes \cdots \otimes \begin{bmatrix} 1 \\ -e^{\jmath 2^{P-p-1}\varpi(0)} \end{bmatrix}. \quad (4.67)$$

Thanks to (4.66), we can express the WNG and the power beampattern of $\mathbf{h}_{\mathrm{H1}}$ as, respectively,

$$\mathcal{W}(\mathbf{h}_{\mathrm{H1}}) = \frac{1}{2^P}\prod_{p=1}^{P}\left|1 - e^{\jmath 3 \times 2^{P-p-1}\varpi(0)}\right|^2$$
$$= \prod_{p=1}^{P}\left\{1 - \cos\left[\frac{3}{2} \times 2^{P-p}\varpi(0)\right]\right\} \quad (4.68)$$

and

$$|\mathcal{B}(\mathbf{h}_{\mathrm{H1}})|^2 = \frac{\prod_{p=1}^{P}\left|1 - e^{3 \times \jmath 2^{P-p-1}\varpi(\theta)}\right|^2}{\prod_{p=1}^{P}\left|1 - e^{3 \times \jmath 2^{P-p-1}\varpi(0)}\right|^2}$$
$$= \frac{\prod_{p=1}^{P}\left\{1 - \cos\left[\frac{3}{2} \times 2^{P-p}\varpi(\theta)\right]\right\}}{\prod_{p=1}^{P}\left\{1 - \cos\left[\frac{3}{2} \times 2^{P-p}\varpi(0)\right]\right\}}. \quad (4.69)$$

Finally, the DF is

$$\mathcal{D}(\mathbf{h}_{\mathrm{H1}}) = \frac{1}{\mathbf{h}_{\mathrm{H1}}^H \mathbf{\Gamma} \mathbf{h}_{\mathrm{H1}}}. \quad (4.70)$$

Figure 4.6 displays the directivity patterns of the first-, second-, and third-order hypercardioid of the first kind for $f = 0.5$ kHz and $\delta = 5$ mm. Figure 4.7 shows plots of the DFs, WNGs, and FBRs of the hypercardioids as a function of frequency. We observe that for low frequencies the DF and FBR increase as we increase the order of the hypercardioid, but the WNG decreases.

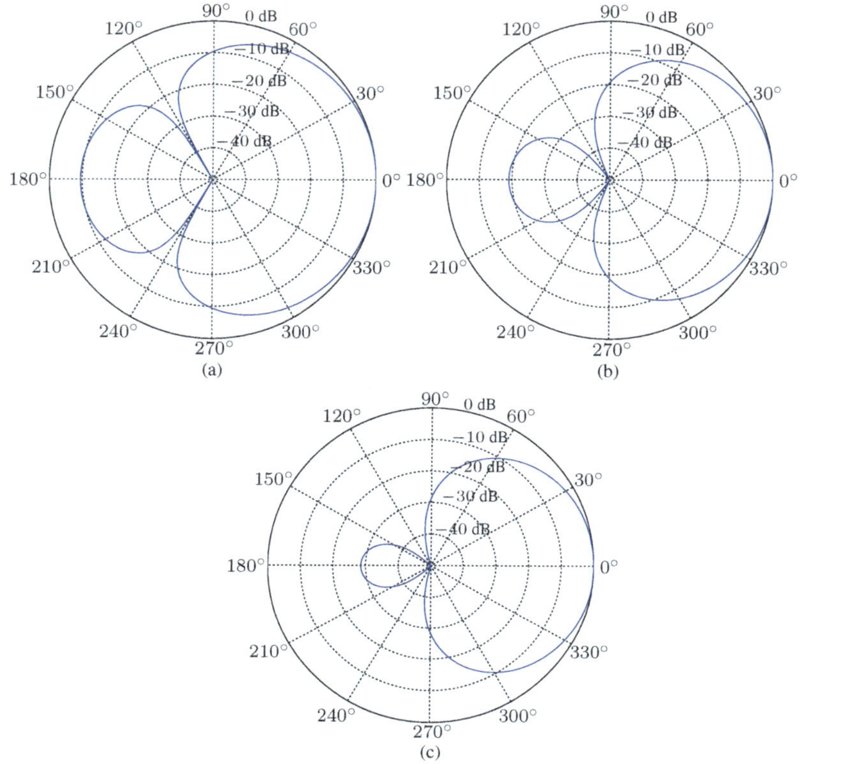

Fig. 4.6 Beampatterns of the first hypercardioid, $\mathbf{h}_{\mathrm{H}1}$, of order P for $f = 0.5$ kHz and $\delta = 5$ mm: (a) $P = 1$, (b) $P = 2$, and (c) $P = 3$.

A more reasonable way to derive the hypercardioid is by maximizing all the DFs, $\mathcal{D}_p(\mathbf{h}_p)$, $p = 1, 2, \ldots, P$. We get

$$\mathbf{h}_p = \frac{\mathbf{\Gamma}_p^{-1}\mathbf{d}_{p,0}}{\mathbf{d}_{p,0}^H\mathbf{\Gamma}_p^{-1}\mathbf{d}_{p,0}}. \tag{4.71}$$

As a result, the second hypercardioid of, roughly, order P that we propose is

$$\mathbf{h}_{\mathrm{H}2} = \mathbf{h}_1 \otimes \mathbf{h}_2 \otimes \cdots \otimes \mathbf{h}_P. \tag{4.72}$$

To fully maximize the DF in (4.31), we need to derive an iterative algorithm. To simplify the presentation, let us take $P = 3$. From the expressions:

$$\mathbf{h}_i \otimes \mathbf{h}_j = (\mathbf{h}_i \otimes \mathbf{I}_2)\,\mathbf{h}_j \tag{4.73}$$

$$= (\mathbf{I}_2 \otimes \mathbf{h}_j)\,\mathbf{h}_i, \tag{4.74}$$

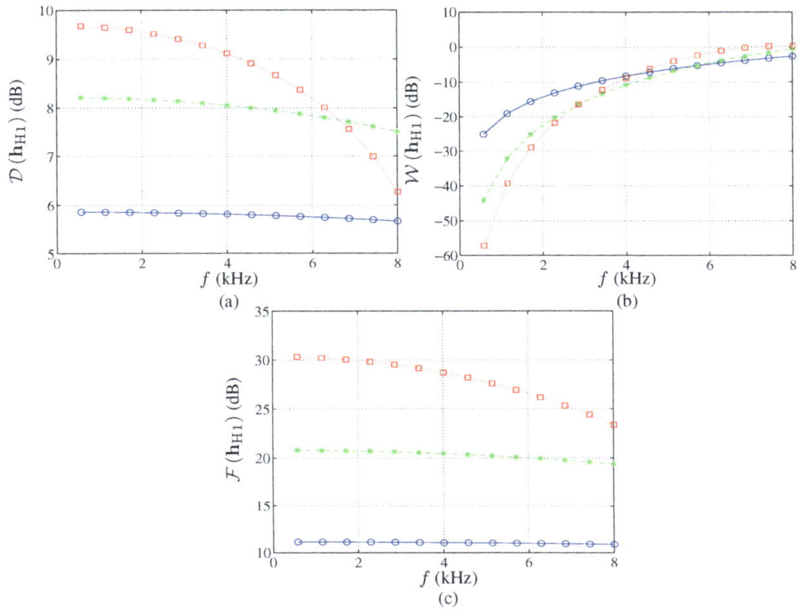

Fig. 4.7 Performance of the first hypercardioid, $\mathbf{h}_{\mathrm{H1}}$, of order P as a function of frequency for $\delta = 5$ mm and several values of P: $P = 1$ (solid line with circles), $P = 2$ (dashed line with asterisks), and $P = 3$ (dotted line with squares). (a) DF, (b) WNG, and (c) FBR.

where $\mathbf{I}_2$ is the 2×2 identity matrix, we deduce the following relationships:

$$\mathbf{h}_1 \otimes \mathbf{h}_2 \otimes \mathbf{h}_3 = (\mathbf{I}_4 \otimes \mathbf{h}_3)(\mathbf{I}_2 \otimes \mathbf{h}_2)\mathbf{h}_1 \tag{4.75}$$

$$= (\mathbf{h}_1 \otimes \mathbf{I}_4)(\mathbf{I}_2 \otimes \mathbf{h}_3)\mathbf{h}_2 \tag{4.76}$$

$$= (\mathbf{h}_1 \otimes \mathbf{I}_4)(\mathbf{h}_2 \otimes \mathbf{I}_2)\mathbf{h}_3, \tag{4.77}$$

where $\mathbf{I}_4$ is the 4×4 identity matrix. Therefore, when $\mathbf{h}_2$ and $\mathbf{h}_3$ are fixed and distortionless, we write the DF as

$$\mathcal{D}(\mathbf{h}_1|\mathbf{h}_2, \mathbf{h}_3) = \frac{\left|\mathbf{h}_1^H \mathbf{d}_{1,0}\right|^2}{\mathbf{h}_1^H \mathbf{\Gamma}_{\mathbf{h}_{2,3}} \mathbf{h}_1}, \tag{4.78}$$

where

$$\mathbf{\Gamma}_{\mathbf{h}_{2,3}} = \left[(\mathbf{I}_4 \otimes \mathbf{h}_3)(\mathbf{I}_2 \otimes \mathbf{h}_2)\right]^H \mathbf{\Gamma} (\mathbf{I}_4 \otimes \mathbf{h}_3)(\mathbf{I}_2 \otimes \mathbf{h}_2). \tag{4.79}$$

When $\mathbf{h}_1$ and $\mathbf{h}_3$ are fixed and distortionless, we write the DF as

$$\mathcal{D}(\mathbf{h}_2|\mathbf{h}_1, \mathbf{h}_3) = \frac{\left|\mathbf{h}_2^H \mathbf{d}_{2,0}\right|^2}{\mathbf{h}_2^H \mathbf{\Gamma}_{\mathbf{h}_{1,3}} \mathbf{h}_2}, \tag{4.80}$$

where

$$\boldsymbol{\Gamma}_{\mathbf{h}_{1,3}} = [(\mathbf{h}_1 \otimes \mathbf{I}_4)(\mathbf{I}_2 \otimes \mathbf{h}_3)]^H \boldsymbol{\Gamma} (\mathbf{h}_1 \otimes \mathbf{I}_4)(\mathbf{I}_2 \otimes \mathbf{h}_3). \tag{4.81}$$

In the same way, when $\mathbf{h}_1$ and $\mathbf{h}_2$ are fixed and distortionless, we write the DF as

$$\mathcal{D}(\mathbf{h}_3|\mathbf{h}_1, \mathbf{h}_2) = \frac{\left|\mathbf{h}_3^H \mathbf{d}_{3,0}\right|^2}{\mathbf{h}_3^H \boldsymbol{\Gamma}_{\mathbf{h}_{1,2}} \mathbf{h}_3}, \tag{4.82}$$

where

$$\boldsymbol{\Gamma}_{\mathbf{h}_{1,2}} = [(\mathbf{h}_1 \otimes \mathbf{I}_4)(\mathbf{h}_2 \otimes \mathbf{I}_2)]^H \boldsymbol{\Gamma} (\mathbf{h}_1 \otimes \mathbf{I}_4)(\mathbf{h}_2 \otimes \mathbf{I}_2). \tag{4.83}$$

At iteration 0, we may take

$$\mathbf{h}_2^{(0)} = \frac{\boldsymbol{\Gamma}_2^{-1} \mathbf{d}_{2,0}}{\mathbf{d}_{2,0}^H \boldsymbol{\Gamma}_2^{-1} \mathbf{d}_{2,0}} \tag{4.84}$$

and

$$\mathbf{h}_3^{(0)} = \frac{\boldsymbol{\Gamma}_3^{-1} \mathbf{d}_{3,0}}{\mathbf{d}_{3,0}^H \boldsymbol{\Gamma}_3^{-1} \mathbf{d}_{3,0}}. \tag{4.85}$$

Substituting $\mathbf{h}_2^{(0)}$ and $\mathbf{h}_3^{(0)}$ into (4.79), we get

$$\boldsymbol{\Gamma}_{\mathbf{h}_{2,3}^{(0)}} = \left[\left(\mathbf{I}_4 \otimes \mathbf{h}_3^{(0)}\right)\left(\mathbf{I}_2 \otimes \mathbf{h}_2^{(0)}\right)\right]^H \boldsymbol{\Gamma} \left(\mathbf{I}_4 \otimes \mathbf{h}_3^{(0)}\right)\left(\mathbf{I}_2 \otimes \mathbf{h}_2^{(0)}\right). \tag{4.86}$$

Now, plugging this expression into the DF in (4.78), we obtain at iteration 1:

$$\mathcal{D}\left(\mathbf{h}_1^{(1)}|\mathbf{h}_2^{(0)}, \mathbf{h}_3^{(0)}\right) = \frac{\left|\left(\mathbf{h}_1^{(1)}\right)^H \mathbf{d}_{1,0}\right|^2}{\left(\mathbf{h}_1^{(1)}\right)^H \boldsymbol{\Gamma}_{\mathbf{h}_{2,3}^{(0)}} \mathbf{h}_1^{(1)}}. \tag{4.87}$$

The maximization of $\mathcal{D}\left(\mathbf{h}_1^{(1)}|\mathbf{h}_2^{(0)}, \mathbf{h}_3^{(0)}\right)$ with respect to $\mathbf{h}_1^{(1)}$ gives

$$\mathbf{h}_1^{(1)} = \frac{\boldsymbol{\Gamma}_{\mathbf{h}_{2,3}^{(0)}}^{-1} \mathbf{d}_{1,0}}{\mathbf{d}_{1,0}^H \boldsymbol{\Gamma}_{\mathbf{h}_{2,3}^{(0)}}^{-1} \mathbf{d}_{1,0}}. \tag{4.88}$$

Using $\mathbf{h}_1^{(1)}$ and $\mathbf{h}_3^{(0)}$ in (4.81), we get

$$\boldsymbol{\Gamma}_{\mathbf{h}_{1,3}^{(1)}} = \left[\left(\mathbf{h}_1^{(1)} \otimes \mathbf{I}_4\right)\left(\mathbf{I}_2 \otimes \mathbf{h}_3^{(0)}\right)\right]^H \boldsymbol{\Gamma} \left(\mathbf{h}_1^{(1)} \otimes \mathbf{I}_4\right)\left(\mathbf{I}_2 \otimes \mathbf{h}_3^{(0)}\right). \tag{4.89}$$

As a consequence, the DF in (4.80) is

$$\mathcal{D}\left(\mathbf{h}_2^{(1)}|\mathbf{h}_1^{(1)},\mathbf{h}_3^{(0)}\right)=\frac{\left|\left(\mathbf{h}_2^{(1)}\right)^H\mathbf{d}_{2,0}\right|^2}{\left(\mathbf{h}_2^{(1)}\right)^H\mathbf{\Gamma}_{\mathbf{h}_{1,3}^{(1)}}\mathbf{h}_2^{(1)}},\tag{4.90}$$

whose maximization with respect to $\mathbf{h}_2^{(1)}$ gives

$$\mathbf{h}_2^{(1)}=\frac{\mathbf{\Gamma}_{\mathbf{h}_{1,3}^{(1)}}^{-1}\mathbf{d}_{2,0}}{\mathbf{d}_{2,0}^H\mathbf{\Gamma}_{\mathbf{h}_{1,3}^{(1)}}^{-1}\mathbf{d}_{2,0}}.\tag{4.91}$$

Finally, the substitution of $\mathbf{h}_1^{(1)}$ and $\mathbf{h}_2^{(1)}$ into (4.83) leads to

$$\mathbf{\Gamma}_{\mathbf{h}_{1,2}^{(1)}}=\left[\left(\mathbf{h}_1^{(1)}\otimes\mathbf{I}_4\right)\left(\mathbf{h}_2^{(1)}\otimes\mathbf{I}_2\right)\right]^H\mathbf{\Gamma}\left(\mathbf{h}_1^{(1)}\otimes\mathbf{I}_4\right)\left(\mathbf{h}_2^{(1)}\otimes\mathbf{I}_2\right),\tag{4.92}$$

then to the DF in (4.82):

$$\mathcal{D}\left(\mathbf{h}_3^{(1)}|\mathbf{h}_1^{(1)},\mathbf{h}_2^{(1)}\right)=\frac{\left|\left(\mathbf{h}_3^{(1)}\right)^H\mathbf{d}_{3,0}\right|^2}{\left(\mathbf{h}_3^{(1)}\right)^H\mathbf{\Gamma}_{\mathbf{h}_{1,2}^{(1)}}\mathbf{h}_3^{(1)}},\tag{4.93}$$

and whose maximization gives

$$\mathbf{h}_3^{(1)}=\frac{\mathbf{\Gamma}_{\mathbf{h}_{1,2}^{(1)}}^{-1}\mathbf{d}_{3,0}}{\mathbf{d}_{3,0}^H\mathbf{\Gamma}_{\mathbf{h}_{1,2}^{(1)}}^{-1}\mathbf{d}_{3,0}}.\tag{4.94}$$

Continuing the iterations up to the iteration n, we easily get for the first filter:

$$\mathbf{h}_1^{(n)}=\frac{\mathbf{\Gamma}_{\mathbf{h}_{2,3}^{(n-1)}}^{-1}\mathbf{d}_{1,0}}{\mathbf{d}_{1,0}^H\mathbf{\Gamma}_{\mathbf{h}_{2,3}^{(n-1)}}^{-1}\mathbf{d}_{1,0}},\tag{4.95}$$

with

$$\mathbf{\Gamma}_{\mathbf{h}_{2,3}^{(n-1)}}=\left[\left(\mathbf{I}_4\otimes\mathbf{h}_3^{(n-1)}\right)\left(\mathbf{I}_2\otimes\mathbf{h}_2^{(n-1)}\right)\right]^H\mathbf{\Gamma}\left(\mathbf{I}_4\otimes\mathbf{h}_3^{(n-1)}\right)\left(\mathbf{I}_2\otimes\mathbf{h}_2^{(n-1)}\right),\tag{4.96}$$

for the second filter:

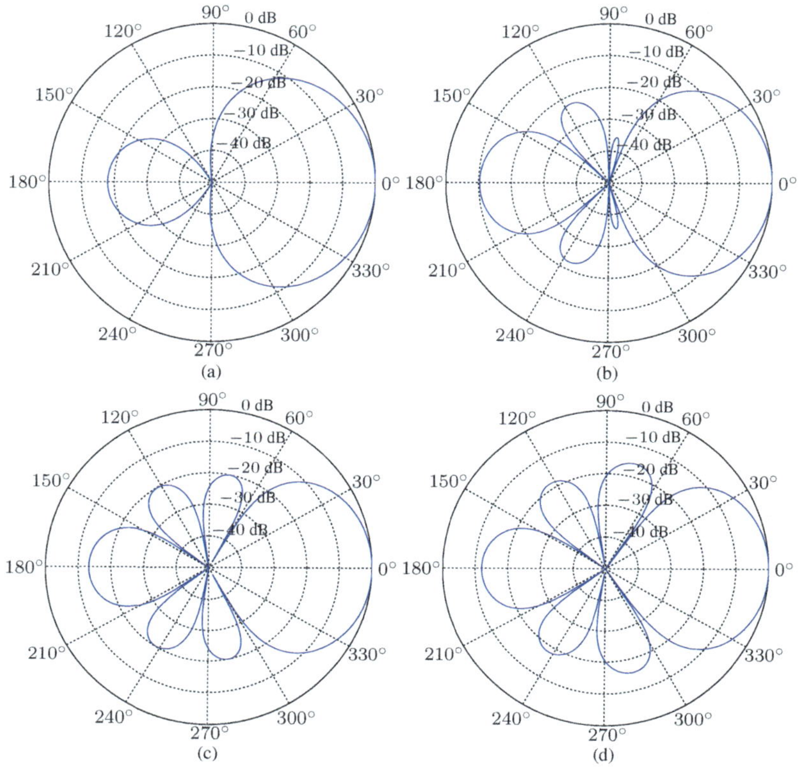

Fig. 4.8 Beampatterns of the third hypercardioid, $\mathbf{h}_{\mathrm{H3}}^{(n)}$, of order $P = 3$ for $f = 0.5$ kHz and $\delta = 5$ mm, obtained at the iteration n: (a) $n = 0$ (the second hypercardioid), (b) $n = 1$, (c) $n = 2$, and (d) $n = 3$.

$$\mathbf{h}_2^{(n)} = \frac{\mathbf{\Gamma}_{\mathbf{h}_{1,3}^{(n)}}^{-1} \mathbf{d}_{2,0}}{\mathbf{d}_{2,0}^H \mathbf{\Gamma}_{\mathbf{h}_{1,3}^{(n)}}^{-1} \mathbf{d}_{2,0}}, \tag{4.97}$$

with

$$\mathbf{\Gamma}_{\mathbf{h}_{1,3}^{(n)}} = \left[\left(\mathbf{h}_1^{(n)} \otimes \mathbf{I}_4 \right) \left(\mathbf{I}_2 \otimes \mathbf{h}_3^{(n-1)} \right) \right]^H \mathbf{\Gamma} \left(\mathbf{h}_1^{(n)} \otimes \mathbf{I}_4 \right) \left(\mathbf{I}_2 \otimes \mathbf{h}_3^{(n-1)} \right), \tag{4.98}$$

and for the third filter:

$$\mathbf{h}_3^{(n)} = \frac{\mathbf{\Gamma}_{\mathbf{h}_{1,2}^{(n)}}^{-1} \mathbf{d}_{3,0}}{\mathbf{d}_{3,0}^H \mathbf{\Gamma}_{\mathbf{h}_{1,2}^{(n)}}^{-1} \mathbf{d}_{3,0}}, \tag{4.99}$$

with

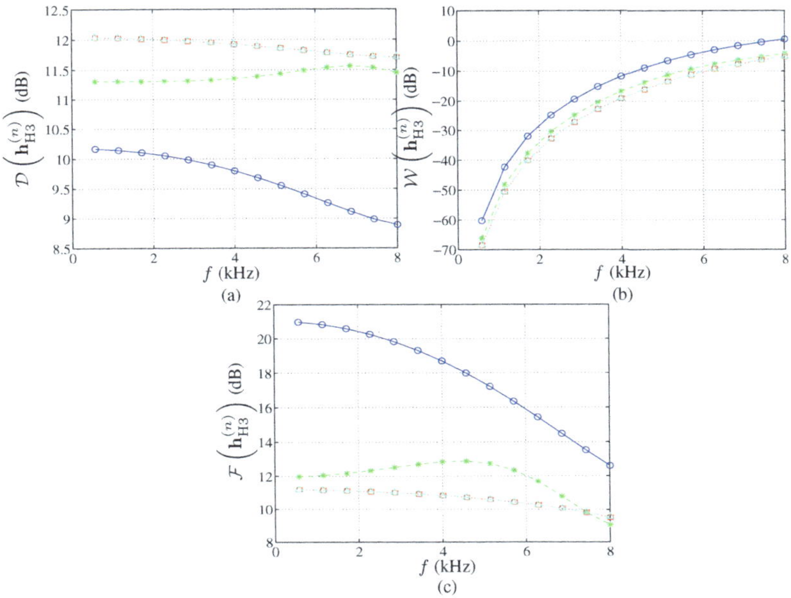

Fig. 4.9 Performance of the third hypercardioid, $\mathbf{h}_{\mathrm{H3}}^{(n)}$, of order $P = 3$ as a function of frequency for $\delta = 5$ mm and several values of n: $n = 0$ (solid line with circles), $n = 1$ (dashed line with asterisks), $n = 3$ (dotted line with squares), and $n = 10$ (dash-dot line with triangles). (a) DF, (b) WNG, and (c) FBR.

$$\boldsymbol{\Gamma}_{\mathbf{h}_{1,2}^{(n)}} = \left[\left(\mathbf{h}_1^{(n)} \otimes \mathbf{I}_4\right)\left(\mathbf{h}_2^{(n)} \otimes \mathbf{I}_2\right)\right]^H \boldsymbol{\Gamma}\left(\mathbf{h}_1^{(n)} \otimes \mathbf{I}_4\right)\left(\mathbf{h}_2^{(n)} \otimes \mathbf{I}_2\right). \quad (4.100)$$

Therefore, the third and last proposed hypercardioid is at iteration n:

$$\mathbf{h}_{\mathrm{H3}}^{(n)} = \mathbf{h}_1^{(n)} \otimes \mathbf{h}_2^{(n)} \otimes \mathbf{h}_3^{(n)}. \quad (4.101)$$

Figure 4.8 displays the directivity patterns of the third-order hypercardioid of the third kind for $f = 0.5$ kHz and $\delta = 5$ mm, obtained at the iteration n for several values of n. Figure 4.9 shows plots of the DFs, WNGs, and FBRs of the hypercardioids as a function of frequency. The iteration $n = 0$ corresponds to the third-order hypercardioid of the second kind. We observe that the DF of the hypercardioid of the third kind is larger than that of the second kind, and the DF of the hypercardioid of the second kind is larger than that of the first kind (compare Figs 4.7 and 4.9). Furthermore, the DF of the hypercardioid of the third kind increases at each iteration, and roughly converges after three iterations, while the WNG and FBR decrease at each iteration.

4.4.5 Supercardioid

The first-order supercardioid has a one at the angle 0 and a null at the angle $3\pi/4$. By simple extension, we can derive a kind of supercardioid of, roughly, order P, as we did for the dipole or cardioid. Therefore, our linear system of two equations is

$$\begin{bmatrix} 1 & e^{\jmath 2^{P-p}\varpi(0)} \\ 1 & e^{\jmath 2^{P-p}\varpi(3\pi/4)} \end{bmatrix} \mathbf{h}_p = \begin{bmatrix} 1 & e^{\jmath 2^{P-p}\varpi(0)} \\ 1 & e^{-\jmath\sqrt{2}\times 2^{P-p-1}\varpi(0)} \end{bmatrix} \mathbf{h}_p = \begin{bmatrix} 1 \\ 0 \end{bmatrix}. \quad (4.102)$$

The solution of the above system is

$$\begin{aligned} \mathbf{h}_p &= \frac{1}{e^{-\jmath\sqrt{2}\times 2^{P-p-1}\varpi(0)} - e^{\jmath 2^{P-p}\varpi(0)}} \begin{bmatrix} e^{-\jmath\sqrt{2}\times 2^{P-p-1}\varpi(0)} \\ -1 \end{bmatrix} \\ &= \frac{1}{1 - e^{\jmath\varphi 2^{P-p-1}\varpi(0)}} \begin{bmatrix} 1 \\ -e^{\jmath\sqrt{2}\times 2^{P-p-1}\varpi(0)} \end{bmatrix}, \end{aligned} \quad (4.103)$$

where $\varphi = 2+\sqrt{2}$. Therefore, the first kind of supercardioid of, roughly, order P that we propose in this subsection is

$$\begin{aligned} \mathbf{h}_{\mathrm{S1}} = &\frac{1}{\prod_{p=1}^{P}\left[1 - e^{\jmath\varphi 2^{P-p-1}\varpi(0)}\right]} \times \\ &\begin{bmatrix} 1 \\ -e^{\jmath\sqrt{2}\times 2^{P-p-1}\varpi(0)} \end{bmatrix} \otimes \cdots \otimes \begin{bmatrix} 1 \\ -e^{\jmath\sqrt{2}\times 2^{P-p-1}\varpi(0)} \end{bmatrix}. \end{aligned} \quad (4.104)$$

Thanks to (4.103), we can express the WNG and the power beampattern of $\mathbf{h}_{\mathrm{S1}}$ as, respectively,

$$\begin{aligned} \mathcal{W}(\mathbf{h}_{\mathrm{S1}}) &= \frac{1}{2^P}\prod_{p=1}^{P}\left|1 - e^{\jmath\varphi 2^{P-p-1}\varpi(0)}\right|^2 \\ &= \prod_{p=1}^{P}\left\{1 - \cos\left[\frac{\varphi}{2}\times 2^{P-p}\varpi(0)\right]\right\} \end{aligned} \quad (4.105)$$

and

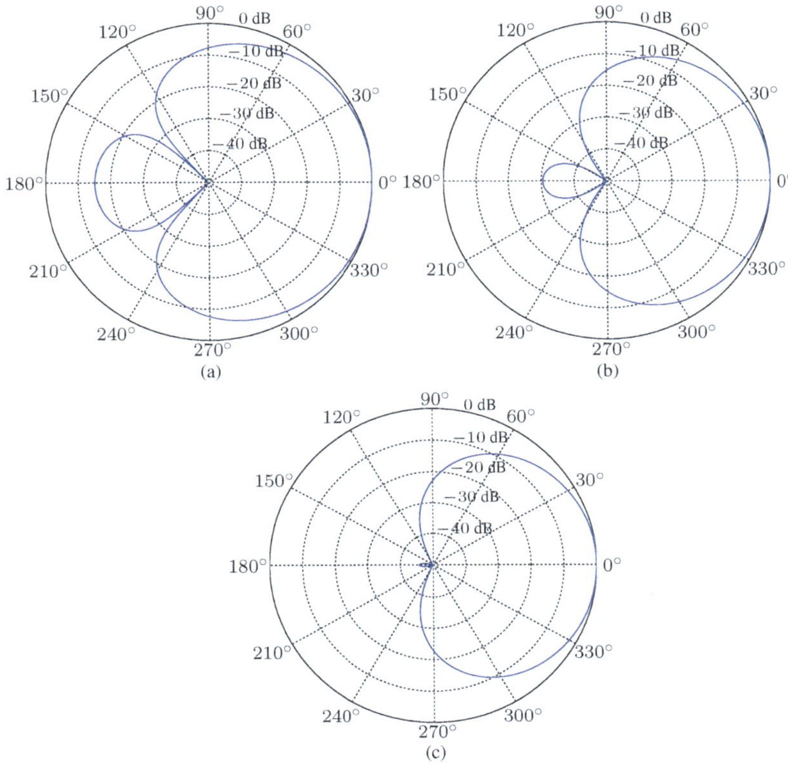

Fig. 4.10 Beampatterns of the first supercardioid, $\mathbf{h}_{S1}$, of order P for $f = 0.5$ kHz and $\delta = 5$ mm: (a) $P = 1$, (b) $P = 2$, and (c) $P = 3$.

$$
\begin{aligned}
|\mathcal{B}(\mathbf{h}_{S1})|^2 &= \frac{\prod_{p=1}^{P}\left|1 - e^{\varphi \jmath 2^{P-p-1}\varpi(\theta)}\right|^2}{\prod_{p=1}^{P}\left|1 - e^{\varphi \jmath 2^{P-p-1}\varpi(0)}\right|^2} \\
&= \frac{\prod_{p=1}^{P}\left\{1 - \cos\left[\frac{\varphi}{2} \times 2^{P-p}\varpi(\theta)\right]\right\}}{\prod_{p=1}^{P}\left\{1 - \cos\left[\frac{\varphi}{2} \times 2^{P-p}\varpi(0)\right]\right\}}.
\end{aligned}
\tag{4.106}
$$

Finally, the DF is

$$
\mathcal{D}(\mathbf{h}_{S1}) = \frac{1}{\mathbf{h}_{S1}^H \boldsymbol{\Gamma} \mathbf{h}_{S1}}. \tag{4.107}
$$

Figure 4.10 displays the directivity patterns of the first-, second- and third-order supercardioid of the first kind for $f = 0.5$ kHz and $\delta = 5$ mm. Figure 4.11 shows plots of the DFs, WNGs, and FBRs of the supercardioids as a function of frequency. We observe that at low frequencies the DF and

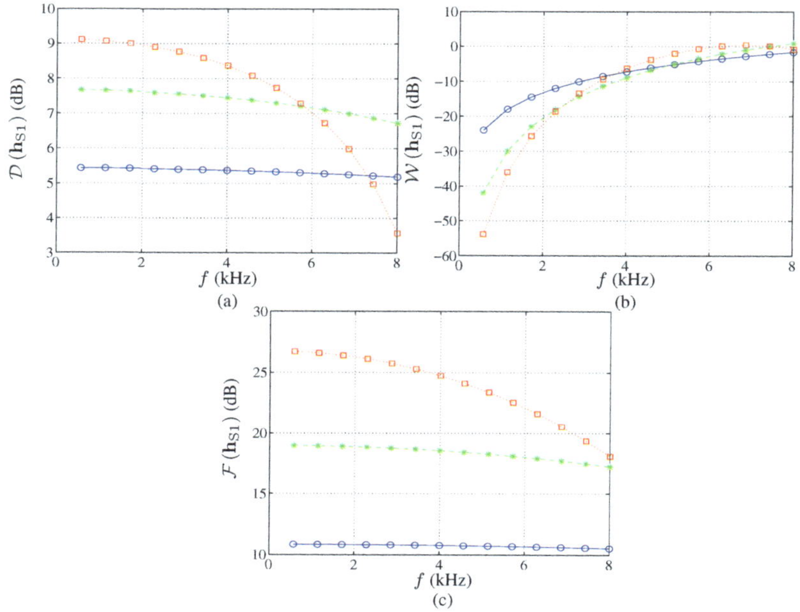

Fig. 4.11 Performance of the first supercardioid, $\mathbf{h}_{S1}$, of order P as a function of frequency for $\delta = 5$ mm and several values of P: $P = 1$ (solid line with circles), $P = 2$ (dashed line with asterisks), and $P = 3$ (dotted line with squares). (a) DF, (b) WNG, and (c) FBR.

FBR increase as we increase the order of the supercardioid, but the WNG decreases.

A more reasonable way to derive the supercardioid is by maximizing all the FBRs, $\mathcal{F}_p(\mathbf{h}_p)$, $p = 1, 2, \ldots, P$. Let $\mathbf{t}_p$ be the eigenvector corresponding to the maximum eigenvalue of $\mathbf{\Gamma}_{b,p}^{-1}\mathbf{\Gamma}_{f,p}$. Then, it is straightforward to see that

$$\mathbf{h}_p = \frac{\mathbf{t}_p}{\mathbf{d}_{p,0}^H \mathbf{t}_p} \tag{4.108}$$

maximizes $\mathcal{F}_p(\mathbf{h}_p)$. As a consequence, the second supercardioid of, roughly, order P that we propose is

$$\mathbf{h}_{S2} = \mathbf{h}_1 \otimes \mathbf{h}_2 \otimes \cdots \otimes \mathbf{h}_P. \tag{4.109}$$

To fully maximize the FBR in (4.39), we need to derive an iterative algorithm as we proposed to do for the maximization of the DF. To simplify the presentation, let us take $P = 3$. When $\mathbf{h}_2$ and $\mathbf{h}_3$ are fixed, we write the FBR as

$$\mathcal{F}(\mathbf{h}_1|\mathbf{h}_2, \mathbf{h}_3) = \frac{\mathbf{h}_1^H \mathbf{\Gamma}_{\mathrm{f},\mathbf{h}_{2,3}} \mathbf{h}_1}{\mathbf{h}_1^H \mathbf{\Gamma}_{\mathrm{b},\mathbf{h}_{2,3}} \mathbf{h}_1}, \tag{4.110}$$

where

$$\mathbf{\Gamma}_{\mathrm{f},\mathbf{h}_{2,3}} = \left[(\mathbf{I}_4 \otimes \mathbf{h}_3)(\mathbf{I}_2 \otimes \mathbf{h}_2)\right]^H \mathbf{\Gamma}_{\mathrm{f}} (\mathbf{I}_4 \otimes \mathbf{h}_3)(\mathbf{I}_2 \otimes \mathbf{h}_2), \tag{4.111}$$

$$\mathbf{\Gamma}_{\mathrm{b},\mathbf{h}_{2,3}} = \left[(\mathbf{I}_4 \otimes \mathbf{h}_3)(\mathbf{I}_2 \otimes \mathbf{h}_2)\right]^H \mathbf{\Gamma}_{\mathrm{b}} (\mathbf{I}_4 \otimes \mathbf{h}_3)(\mathbf{I}_2 \otimes \mathbf{h}_2). \tag{4.112}$$

When $\mathbf{h}_1$ and $\mathbf{h}_3$ are fixed, we write the FBR as

$$\mathcal{F}(\mathbf{h}_2|\mathbf{h}_1, \mathbf{h}_3) = \frac{\mathbf{h}_2^H \mathbf{\Gamma}_{\mathrm{f},\mathbf{h}_{1,3}} \mathbf{h}_2}{\mathbf{h}_2^H \mathbf{\Gamma}_{\mathrm{b},\mathbf{h}_{1,3}} \mathbf{h}_2}, \tag{4.113}$$

where

$$\mathbf{\Gamma}_{\mathrm{f},\mathbf{h}_{1,3}} = \left[(\mathbf{h}_1 \otimes \mathbf{I}_4)(\mathbf{I}_2 \otimes \mathbf{h}_3)\right]^H \mathbf{\Gamma}_{\mathrm{f}} (\mathbf{h}_1 \otimes \mathbf{I}_4)(\mathbf{I}_2 \otimes \mathbf{h}_3), \tag{4.114}$$

$$\mathbf{\Gamma}_{\mathrm{b},\mathbf{h}_{1,3}} = \left[(\mathbf{h}_1 \otimes \mathbf{I}_4)(\mathbf{I}_2 \otimes \mathbf{h}_3)\right]^H \mathbf{\Gamma}_{\mathrm{b}} (\mathbf{h}_1 \otimes \mathbf{I}_4)(\mathbf{I}_2 \otimes \mathbf{h}_3). \tag{4.115}$$

In the same way, when $\mathbf{h}_1$ and $\mathbf{h}_2$ are fixed, we write the FBR as

$$\mathcal{F}(\mathbf{h}_3|\mathbf{h}_1, \mathbf{h}_2) = \frac{\mathbf{h}_3^H \mathbf{\Gamma}_{\mathrm{f},\mathbf{h}_{1,2}} \mathbf{h}_3}{\mathbf{h}_3^H \mathbf{\Gamma}_{\mathrm{b},\mathbf{h}_{1,2}} \mathbf{h}_3}, \tag{4.116}$$

where

$$\mathbf{\Gamma}_{\mathrm{f},\mathbf{h}_{1,2}} = \left[(\mathbf{h}_1 \otimes \mathbf{I}_4)(\mathbf{h}_2 \otimes \mathbf{I}_2)\right]^H \mathbf{\Gamma}_{\mathrm{f}} (\mathbf{h}_1 \otimes \mathbf{I}_4)(\mathbf{h}_2 \otimes \mathbf{I}_2), \tag{4.117}$$

$$\mathbf{\Gamma}_{\mathrm{b},\mathbf{h}_{1,2}} = \left[(\mathbf{h}_1 \otimes \mathbf{I}_4)(\mathbf{h}_2 \otimes \mathbf{I}_2)\right]^H \mathbf{\Gamma}_{\mathrm{b}} (\mathbf{h}_1 \otimes \mathbf{I}_4)(\mathbf{h}_2 \otimes \mathbf{I}_2). \tag{4.118}$$

At iteration 0, we may take

$$\mathbf{h}_2^{(0)} = \frac{\mathbf{t}_2}{\mathbf{d}_{2,0}^H \mathbf{t}_2} \tag{4.119}$$

and

$$\mathbf{h}_3^{(0)} = \frac{\mathbf{t}_3}{\mathbf{d}_{3,0}^H \mathbf{t}_3}, \tag{4.120}$$

where $\mathbf{t}_2$ and $\mathbf{t}_3$ are the eigenvectors corresponding to the maximum eigenvalues of $\mathbf{\Gamma}_{\mathrm{b},2}^{-1}\mathbf{\Gamma}_{\mathrm{f},2}$ and $\mathbf{\Gamma}_{\mathrm{b},3}^{-1}\mathbf{\Gamma}_{\mathrm{f},3}$, respectively. Substituting $\mathbf{h}_2^{(0)}$ and $\mathbf{h}_3^{(0)}$ into (4.111) and (4.112), we get

$$\mathbf{\Gamma}_{\mathrm{f},\mathbf{h}_{2,3}^{(0)}} = \left[\left(\mathbf{I}_4 \otimes \mathbf{h}_3^{(0)}\right)\left(\mathbf{I}_2 \otimes \mathbf{h}_2^{(0)}\right)\right]^H \mathbf{\Gamma}_{\mathrm{f}} \left(\mathbf{I}_4 \otimes \mathbf{h}_3^{(0)}\right)\left(\mathbf{I}_2 \otimes \mathbf{h}_2^{(0)}\right), \tag{4.121}$$

$$\mathbf{\Gamma}_{\mathrm{b},\mathbf{h}_{2,3}^{(0)}} = \left[\left(\mathbf{I}_4 \otimes \mathbf{h}_3^{(0)}\right)\left(\mathbf{I}_2 \otimes \mathbf{h}_2^{(0)}\right)\right]^H \mathbf{\Gamma}_{\mathrm{b}} \left(\mathbf{I}_4 \otimes \mathbf{h}_3^{(0)}\right)\left(\mathbf{I}_2 \otimes \mathbf{h}_2^{(0)}\right). \tag{4.122}$$

Now, plugging these expressions into the FBR in (4.110), we obtain at iteration 1:

$$\mathcal{F}\left(\mathbf{h}_1^{(1)}|\mathbf{h}_2^{(0)},\mathbf{h}_3^{(0)}\right)=\frac{\left(\mathbf{h}_1^{(1)}\right)^H\boldsymbol{\Gamma}_{\mathrm{f},\mathbf{h}_{2,3}^{(0)}}\mathbf{h}_1^{(1)}}{\left(\mathbf{h}_1^{(1)}\right)^H\boldsymbol{\Gamma}_{\mathrm{b},\mathbf{h}_{2,3}^{(0)}}\mathbf{h}_1^{(1)}}. \quad (4.123)$$

The maximization of $\mathcal{F}\left(\mathbf{h}_1^{(1)}|\mathbf{h}_2^{(0)},\mathbf{h}_3^{(0)}\right)$ with respect to $\mathbf{h}_1^{(1)}$ gives

$$\mathbf{h}_1^{(1)}=\frac{\mathbf{t}_1^{(0)}}{\mathbf{d}_{1,0}^H\mathbf{t}_1^{(0)}}, \quad (4.124)$$

where $\mathbf{t}_1^{(0)}$ is the eigenvector corresponding to the maximum eigenvalue of $\boldsymbol{\Gamma}_{\mathrm{b},\mathbf{h}_{2,3}^{(0)}}^{-1}\boldsymbol{\Gamma}_{\mathrm{f},\mathbf{h}_{2,3}^{(0)}}$. Using $\mathbf{h}_1^{(1)}$ and $\mathbf{h}_3^{(0)}$ in (4.114) and (4.115), we get

$$\boldsymbol{\Gamma}_{\mathrm{f},\mathbf{h}_{1,3}^{(1)}}=\left[\left(\mathbf{h}_1^{(1)}\otimes\mathbf{I}_4\right)\left(\mathbf{I}_2\otimes\mathbf{h}_3^{(0)}\right)\right]^H\boldsymbol{\Gamma}_{\mathrm{f}}\left(\mathbf{h}_1^{(1)}\otimes\mathbf{I}_4\right)\left(\mathbf{I}_2\otimes\mathbf{h}_3^{(0)}\right), \quad (4.125)$$

$$\boldsymbol{\Gamma}_{\mathrm{b},\mathbf{h}_{1,3}^{(1)}}=\left[\left(\mathbf{h}_1^{(1)}\otimes\mathbf{I}_4\right)\left(\mathbf{I}_2\otimes\mathbf{h}_3^{(0)}\right)\right]^H\boldsymbol{\Gamma}_{\mathrm{b}}\left(\mathbf{h}_1^{(1)}\otimes\mathbf{I}_4\right)\left(\mathbf{I}_2\otimes\mathbf{h}_3^{(0)}\right). \quad (4.126)$$

As a consequence, the FBR in (4.113) is

$$\mathcal{F}\left(\mathbf{h}_2^{(1)}|\mathbf{h}_1^{(1)},\mathbf{h}_3^{(0)}\right)=\frac{\left(\mathbf{h}_2^{(1)}\right)^H\boldsymbol{\Gamma}_{\mathrm{f},\mathbf{h}_{1,3}^{(1)}}\mathbf{h}_2^{(1)}}{\left(\mathbf{h}_2^{(1)}\right)^H\boldsymbol{\Gamma}_{\mathrm{b},\mathbf{h}_{1,3}^{(1)}}\mathbf{h}_2^{(1)}}, \quad (4.127)$$

whose maximization with respect to $\mathbf{h}_2^{(1)}$ gives

$$\mathbf{h}_2^{(1)}=\frac{\mathbf{t}_2^{(1)}}{\mathbf{d}_{2,0}^H\mathbf{t}_2^{(1)}}, \quad (4.128)$$

where $\mathbf{t}_2^{(1)}$ is the eigenvector corresponding to the maximum eigenvalue of $\boldsymbol{\Gamma}_{\mathrm{b},\mathbf{h}_{1,3}^{(1)}}^{-1}\boldsymbol{\Gamma}_{\mathrm{f},\mathbf{h}_{1,3}^{(1)}}$. Finally, the substitution of $\mathbf{h}_1^{(1)}$ and $\mathbf{h}_2^{(1)}$ into (4.117) and (4.118) leads to

$$\boldsymbol{\Gamma}_{\mathrm{f},\mathbf{h}_{1,2}^{(1)}}=\left[\left(\mathbf{h}_1^{(1)}\otimes\mathbf{I}_4\right)\left(\mathbf{h}_2^{(1)}\otimes\mathbf{I}_2\right)\right]^H\boldsymbol{\Gamma}_{\mathrm{f}}\left(\mathbf{h}_1^{(1)}\otimes\mathbf{I}_4\right)\left(\mathbf{h}_2^{(1)}\otimes\mathbf{I}_2\right), \quad (4.129)$$

$$\boldsymbol{\Gamma}_{\mathrm{b},\mathbf{h}_{1,2}^{(1)}}=\left[\left(\mathbf{h}_1^{(1)}\otimes\mathbf{I}_4\right)\left(\mathbf{h}_2^{(1)}\otimes\mathbf{I}_2\right)\right]^H\boldsymbol{\Gamma}_{\mathrm{b}}\left(\mathbf{h}_1^{(1)}\otimes\mathbf{I}_4\right)\left(\mathbf{h}_2^{(1)}\otimes\mathbf{I}_2\right), \quad (4.130)$$

then to the FBR in (4.116):

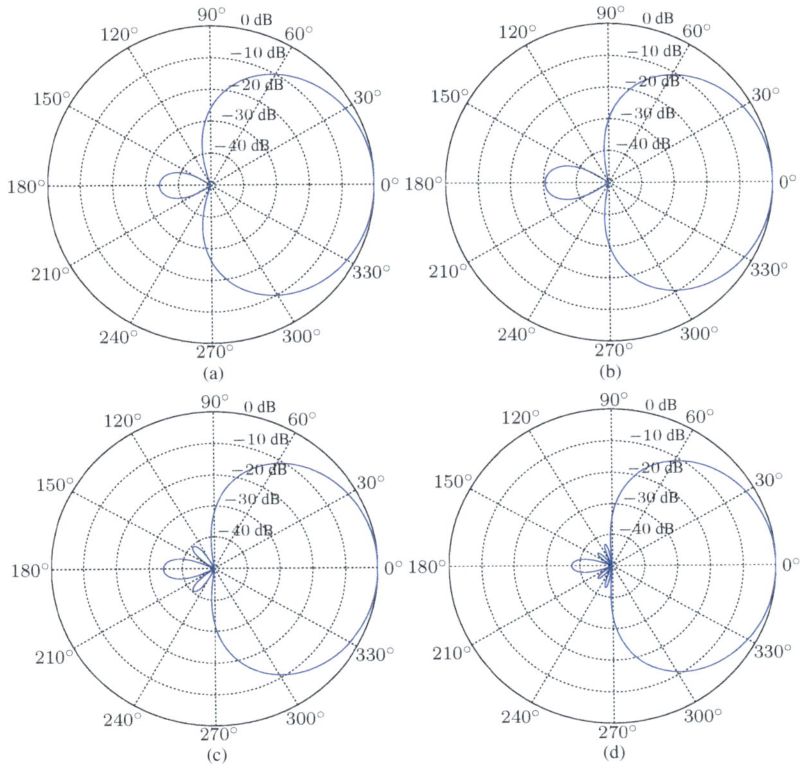

Fig. 4.12 Beampatterns of the third supercardioid, $\mathbf{h}_{S3}^{(n)}$, of order $P = 3$ for $f = 0.5$ kHz and $\delta = 5$ mm, obtained at the iteration n: (a) $n = 0$ (the second supercardioid), (b) $n = 1$, (c) $n = 2$, and (d) $n = 5$.

$$\mathcal{F}\left(\mathbf{h}_3^{(1)}|\mathbf{h}_1^{(1)}, \mathbf{h}_2^{(1)}\right) = \frac{\left(\mathbf{h}_3^{(1)}\right)^H \mathbf{\Gamma}_{\mathrm{f},\mathbf{h}_{1,2}^{(1)}} \mathbf{h}_3^{(1)}}{\left(\mathbf{h}_3^{(1)}\right)^H \mathbf{\Gamma}_{\mathrm{b},\mathbf{h}_{1,2}^{(1)}} \mathbf{h}_3^{(1)}}, \tag{4.131}$$

and whose maximization gives

$$\mathbf{h}_3^{(1)} = \frac{\mathbf{t}_3^{(1)}}{\mathbf{d}_{3,0}^H \mathbf{t}_3^{(1)}}, \tag{4.132}$$

where $\mathbf{t}_3^{(1)}$ is the eigenvector corresponding to the maximum eigenvalue of $\mathbf{\Gamma}_{\mathrm{b},\mathbf{h}_{1,2}^{(1)}}^{-1} \mathbf{\Gamma}_{\mathrm{f},\mathbf{h}_{1,2}^{(1)}}$.

Continuing the iterations up to the iteration n, we easily get for the first filter:

$$\mathbf{h}_1^{(n)} = \frac{\mathbf{t}_1^{(n-1)}}{\mathbf{d}_{1,0}^H \mathbf{t}_1^{(n-1)}}, \tag{4.133}$$

where $\mathbf{t}_1^{(n-1)}$ is the eigenvector corresponding to the maximum eigenvalue of $\mathbf{\Gamma}_{\mathrm{b},\mathbf{h}_{2,3}^{(n-1)}}^{-1} \mathbf{\Gamma}_{\mathrm{f},\mathbf{h}_{2,3}^{(n-1)}}$, with

$$\mathbf{\Gamma}_{\mathrm{f},\mathbf{h}_{2,3}^{(n-1)}} = \left[\left(\mathbf{I}_4 \otimes \mathbf{h}_3^{(n-1)}\right)\left(\mathbf{I}_2 \otimes \mathbf{h}_2^{(n-1)}\right)\right]^H \mathbf{\Gamma}_{\mathrm{f}} \left(\mathbf{I}_4 \otimes \mathbf{h}_3^{(n-1)}\right)\left(\mathbf{I}_2 \otimes \mathbf{h}_2^{(n-1)}\right), \tag{4.134}$$

$$\mathbf{\Gamma}_{\mathrm{b},\mathbf{h}_{2,3}^{(n-1)}} = \left[\left(\mathbf{I}_4 \otimes \mathbf{h}_3^{(n-1)}\right)\left(\mathbf{I}_2 \otimes \mathbf{h}_2^{(n-1)}\right)\right]^H \mathbf{\Gamma}_{\mathrm{b}} \left(\mathbf{I}_4 \otimes \mathbf{h}_3^{(n-1)}\right)\left(\mathbf{I}_2 \otimes \mathbf{h}_2^{(n-1)}\right), \tag{4.135}$$

for the second filter:

$$\mathbf{h}_2^{(n)} = \frac{\mathbf{t}_2^{(n)}}{\mathbf{d}_{2,0}^H \mathbf{t}_2^{(n)}}, \tag{4.136}$$

where $\mathbf{t}_2^{(n)}$ is the eigenvector corresponding to the maximum eigenvalue of $\mathbf{\Gamma}_{\mathrm{b},\mathbf{h}_{1,3}^{(n)}}^{-1} \mathbf{\Gamma}_{\mathrm{f},\mathbf{h}_{1,3}^{(n)}}$, with

$$\mathbf{\Gamma}_{\mathrm{f},\mathbf{h}_{1,3}^{(n)}} = \left[\left(\mathbf{h}_1^{(n)} \otimes \mathbf{I}_4\right)\left(\mathbf{I}_2 \otimes \mathbf{h}_3^{(n-1)}\right)\right]^H \mathbf{\Gamma}_{\mathrm{f}} \left(\mathbf{h}_1^{(n)} \otimes \mathbf{I}_4\right)\left(\mathbf{I}_2 \otimes \mathbf{h}_3^{(n-1)}\right), \tag{4.137}$$

$$\mathbf{\Gamma}_{\mathrm{b},\mathbf{h}_{1,3}^{(n)}} = \left[\left(\mathbf{h}_1^{(n)} \otimes \mathbf{I}_4\right)\left(\mathbf{I}_2 \otimes \mathbf{h}_3^{(n-1)}\right)\right]^H \mathbf{\Gamma}_{\mathrm{b}} \left(\mathbf{h}_1^{(n)} \otimes \mathbf{I}_4\right)\left(\mathbf{I}_2 \otimes \mathbf{h}_3^{(n-1)}\right), \tag{4.138}$$

and for the third filter:

$$\mathbf{h}_3^{(n)} = \frac{\mathbf{t}_3^{(n)}}{\mathbf{d}_{3,0}^H \mathbf{t}_3^{(n)}}, \tag{4.139}$$

where $\mathbf{t}_3^{(n)}$ is the eigenvector corresponding to the maximum eigenvalue of $\mathbf{\Gamma}_{\mathrm{b},\mathbf{h}_{1,2}^{(n)}}^{-1} \mathbf{\Gamma}_{\mathrm{f},\mathbf{h}_{1,2}^{(n)}}$, with

$$\mathbf{\Gamma}_{\mathrm{f},\mathbf{h}_{1,2}^{(n)}} = \left[\left(\mathbf{h}_1^{(n)} \otimes \mathbf{I}_4\right)\left(\mathbf{h}_2^{(n)} \otimes \mathbf{I}_2\right)\right]^H \mathbf{\Gamma}_{\mathrm{f}} \left(\mathbf{h}_1^{(n)} \otimes \mathbf{I}_4\right)\left(\mathbf{h}_2^{(n)} \otimes \mathbf{I}_2\right), \tag{4.140}$$

$$\mathbf{\Gamma}_{\mathrm{b},\mathbf{h}_{1,2}^{(n)}} = \left[\left(\mathbf{h}_1^{(n)} \otimes \mathbf{I}_4\right)\left(\mathbf{h}_2^{(n)} \otimes \mathbf{I}_2\right)\right]^H \mathbf{\Gamma}_{\mathrm{b}} \left(\mathbf{h}_1^{(n)} \otimes \mathbf{I}_4\right)\left(\mathbf{h}_2^{(n)} \otimes \mathbf{I}_2\right). \tag{4.141}$$

Therefore, the third and last proposed supercardioid is at iteration n:

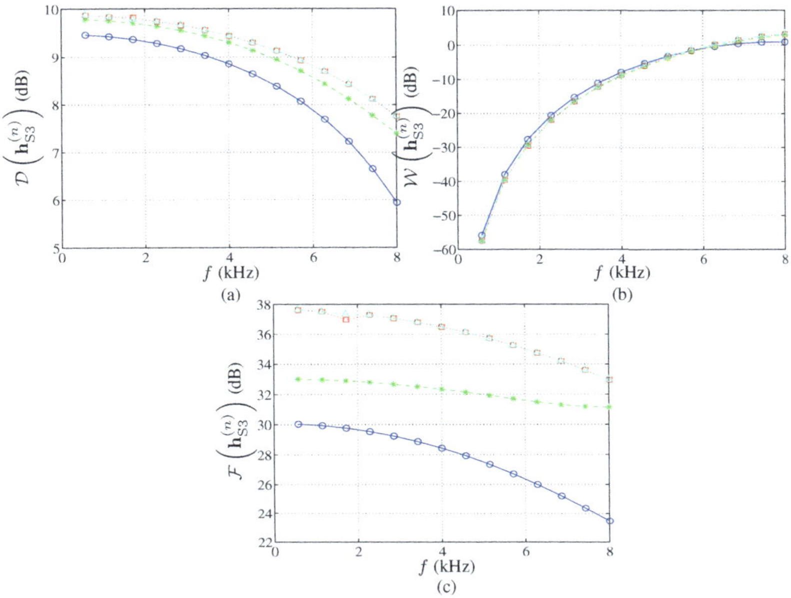

Fig. 4.13 Performance of the third supercardioid, $\mathbf{h}_{S3}^{(n)}$, of order $P = 3$ as a function of frequency for $\delta = 5$ mm and several values of n: $n = 0$ (solid line with circles), $n = 1$ (dashed line with asterisks), $n = 5$ (dotted line with squares), and $n = 10$ (dash-dot line with triangles). (a) DF, (b) WNG, and (c) FBR.

$$\mathbf{h}_{S3}^{(n)} = \mathbf{h}_1^{(n)} \otimes \mathbf{h}_2^{(n)} \otimes \mathbf{h}_3^{(n)}. \tag{4.142}$$

Figure 4.12 displays the directivity patterns of the third-order supercardioid of the third kind for $f = 0.5$ kHz and $\delta = 5$ mm, obtained at the iteration n for several values of n. Figure 4.13 shows plots of the DFs, WNGs, and FBRs of the supercardioids as a function of frequency. The iteration $n = 0$ corresponds to the third-order supercardioid of the second kind. We observe that the FBR of the supercardioid of the third kind is larger than that of the second kind, and the FBR of the supercardioid of the second kind is larger than that of the first kind (compare Figs 4.11 and 4.13). Furthermore, the FBR of the supercardioid of the third kind increases at each iteration, and roughly converges after five iterations, while the DF also increases at each iteration.

References

1. D. H. Johnson and D. E. Dudgeon, *Array Signal Processing: Concepts and Techniques*. Signal Processing Series. Englewood Cliffs, NJ: Prentice-Hall, 1993.

2. H. L. Van Trees, *Optimum Array Processing: Part IV of Detection, Estimation, and Modulation Theory.* New York, NY: John Wiley & Sons, Inc., 2002.
3. G. Zhu and K. Huang, "Analog spatial cancellation for tackling the near-far problem in wirelessly powered communications," *IEEE J. Selected Areas Communications*, vol. 34, pp. 3566–3576, Dec. 2016.
4. J. Benesty, J. Chen, and Y. Huang, *Microphone Array Signal Processing.* Berlin, Germany: Springer-Verlag, 2008.
5. J. Benesty, I. Cohen, and J. Chen, *Fundamentals of Signal Enhancement and Array Signal Processing.* Singapore: Wiley–IEEE Press, 2018.
6. J. Benesty and J. Chen, *Study and Design of Differential Microphone Arrays.* Berlin, Germany: Springer-Verlag, 2012.
7. J. Benesty, J. Chen, and C. Pan, *Fundamentals of Differential Beamforming.* Springer Briefs in Electrical & Computer Engineering, 2016.

Chapter 5
Approach with Nonuniform Linear Arrays

Until now, we have dealt with ULAs but another straightforward geometry extension is the nonuniform linear array (NULA), which is the focus of this chapter. First, we show how from two virtual ULAs we can construct a physical NULA whose associated steering vector is the Kronecker product of the steering vectors associated with the virtual arrays. Then, we explain how Kronecker product beamforming is performed. We give the most important performance measures in this context. Finally, we show how to derive some interesting optimal beamformers.

5.1 Signal Model and Problem Formulation

To describe our model, let us start by considering two different virtual ULAs on the same line denoted 1 and 2. ULA 1 has M_1 microphones and an interelement spacing equal to $M_2\delta$, where δ is some unit spacing, and ULA 2 has M_2 microphones and an interelement spacing equal to $M_1\delta$ (some examples are shown in Fig. 1.3). Now, let us assume that a desired source signal (plane wave), in the farfield, propagates from the azimuth angle, θ, in an anechoic acoustic environment at the speed of sound and impinges on the above described virtual arrays. In this scenario, the corresponding steering vectors (of lengths M_1 and M_2 respectively) are [1]

$$\mathbf{d}_{1,\theta} = \begin{bmatrix} 1 & e^{-\jmath M_2 \varpi(\theta)} & e^{-\jmath 2M_2 \varpi(\theta)} & \cdots & e^{-\jmath (M_1-1)M_2 \varpi(\theta)} \end{bmatrix}^T, \tag{5.1}$$

$$\mathbf{d}_{2,\theta} = \begin{bmatrix} 1 & e^{-\jmath M_1 \varpi(\theta)} & e^{-\jmath 2M_1 \varpi(\theta)} & \cdots & e^{-\jmath (M_2-1)M_1 \varpi(\theta)} \end{bmatrix}^T, \tag{5.2}$$

where

$$\varpi(\theta) = \frac{\omega\delta\cos\theta}{c}. \tag{5.3}$$

J. Benesty et al., *Array Processing*, Springer Topics in Signal Processing 18, https://doi.org/10.1007/978-3-030-15600-8_5

Without loss of generality, we may assume that $M_2 > M_1$ (with $M_1 \geq 2$ and $M_2 \geq 3$), so that $(M_2 - 1)M_1 > (M_1 - 1)M_2$. In this study, we are interested in physical linear arrays whose associated steering vectors are of the form:

$$\begin{aligned}\widetilde{\mathbf{d}}_\theta &= \mathbf{d}_{1,\theta} \otimes \mathbf{d}_{2,\theta} \\ &= \left[\mathbf{d}_{2,\theta}^T \quad e^{-\jmath M_2 \varpi(\theta)} \mathbf{d}_{2,\theta}^T \quad e^{-\jmath 2 M_2 \varpi(\theta)} \mathbf{d}_{2,\theta}^T \quad \cdots \quad e^{-\jmath (M_1 - 1) M_2 \varpi(\theta)} \mathbf{d}_{2,\theta}^T \right]^T .\end{aligned} \quad (5.4)$$

Clearly, $\widetilde{\mathbf{d}}_\theta$ is of length $M_1 M_2$. Furthermore, it will be assumed that M_1 and M_2 are coprime integers, so that all the sensors (whose number is equal to $M_1 M_2$) of the formed linear array are in different positions; however, the mth ($m = 1, 2, \ldots, M_1 M_2$) element of $\widetilde{\mathbf{d}}_\theta$ does not necessarily correspond to the mth array sensor (except for the first and last entries, i.e., $m = 1$ and $m = M_1 M_2$) and this steering vector needs to be reordered appropriately, which is easy to do when M_1 and M_2 are given. As a consequence, we have built a physical NULA with $M_1 M_2$ sensors (in different positions on the same line) whose associated (unordered) steering vector is given in (5.4) and whose aperture is $(2M_1 M_2 - M_1 - M_2)\delta$. Notice that what we propose here is not a coprime array [2] since we have $M_1 M_2$ physical sensors while a coprime array has only $M_1 + M_2 - 1$ elements. Furthermore, the proposed processing is very much different. We only borrow the fact that M_1 and M_2 are coprime integers so that the resulting NULA has non-overlapping elements.

In this chapter, we consider doing beamforming with small values of δ, like in differential [3], [4], [5] or superdirective [6], [7] beamforming, where the main lobe is at the angle $\theta = 0$ (endfire direction) and the desired signal propagates from the same angle. Beamforming under these conditions has the potential to lead to large array gains. Then, our objective is to study beamforming with the proposed NULAs. Different avenues will be investigated.

Since the source propagates from the angle $\theta = 0$, the observation signal vector of length $M_1 M_2$ can be expressed in the frequency domain as [8], [9]

$$\begin{aligned}\widetilde{\mathbf{y}} &= \left[\widetilde{Y}_1 \;\; \widetilde{Y}_2 \;\; \cdots \;\; \widetilde{Y}_{M_1 M_2} \right]^T \\ &= \widetilde{\mathbf{x}} + \widetilde{\mathbf{v}} \\ &= \widetilde{\mathbf{d}}_0 X + \widetilde{\mathbf{v}},\end{aligned} \quad (5.5)$$

where $\widetilde{\mathbf{d}}_0 = \mathbf{d}_{1,0} \otimes \mathbf{d}_{2,0}$ is the steering vector at $\theta = 0$, X is the zero-mean desired source signal, and $\widetilde{\mathbf{v}}$ is the zero-mean additive noise signal vector defined similarly to $\widetilde{\mathbf{y}}$. The covariance matrix of $\widetilde{\mathbf{y}}$ is then

$$\begin{aligned}\mathbf{\Phi}_{\widetilde{\mathbf{y}}} &= E\left(\widetilde{\mathbf{y}} \widetilde{\mathbf{y}}^H \right) \\ &= \phi_X \widetilde{\mathbf{d}}_0 \widetilde{\mathbf{d}}_0^H + \mathbf{\Phi}_{\widetilde{\mathbf{v}}},\end{aligned} \quad (5.6)$$

where $\phi_X = E\left(|X|^2\right)$ is the variance of X and $\mathbf{\Phi}_{\widetilde{\mathbf{v}}} = E\left(\widetilde{\mathbf{v}}\widetilde{\mathbf{v}}^H\right)$ is the covariance matrix of $\widetilde{\mathbf{v}}$. Assuming that the first sensor is the reference, we can express (5.6) as

$$\mathbf{\Phi}_{\widetilde{\mathbf{y}}} = \phi_X \widetilde{\mathbf{d}}_0 \widetilde{\mathbf{d}}_0^H + \phi_{\widetilde{V}_1} \mathbf{\Gamma}_{\widetilde{\mathbf{v}}}, \tag{5.7}$$

where $\phi_{\widetilde{V}_1} = E\left(\left|\widetilde{V}_1\right|^2\right)$ is the variance of the noise at the reference sensor and $\mathbf{\Gamma}_{\widetilde{\mathbf{v}}} = \mathbf{\Phi}_{\widetilde{\mathbf{v}}}/\phi_{\widetilde{V}_1}$ is the pseudo-coherence matrix of the noise. In the case of the spherically isotropic (diffuse) noise field, which will often be assumed here, (5.7) becomes

$$\mathbf{\Phi}_{\widetilde{\mathbf{y}}} = \phi_X \widetilde{\mathbf{d}}_0 \widetilde{\mathbf{d}}_0^H + \phi \widetilde{\mathbf{\Gamma}}, \tag{5.8}$$

where ϕ is the variance of the diffuse noise and

$$\begin{aligned}\widetilde{\mathbf{\Gamma}} &= \frac{1}{2}\int_0^{\pi} \widetilde{\mathbf{d}}_\theta \widetilde{\mathbf{d}}_\theta^H \sin\theta d\theta \qquad (5.9)\\ &= \frac{1}{2}\int_0^{\pi} \left(\mathbf{d}_{1,\theta}\mathbf{d}_{1,\theta}^H\right) \otimes \left(\mathbf{d}_{2,\theta}\mathbf{d}_{2,\theta}^H\right) \sin\theta d\theta.\end{aligned}$$

From the signal model described in this section, we will show next how to perform beamforming with Kronecker product filters.

5.2 Kronecker Product Beamforming

We propose to perform beamforming with the Kronecker product filter of length M_1M_2:

$$\widetilde{\mathbf{h}} = \mathbf{h}_1 \otimes \mathbf{h}_2, \tag{5.10}$$

where $\mathbf{h}_1$ and $\mathbf{h}_2$ are two complex-valued linear filters of lengths M_1 and M_2, respectively. Then, beamforming is performed by applying $\widetilde{\mathbf{h}}$ [as defined in (5.10)] to $\widetilde{\mathbf{y}}$ [from (5.5)]. We get

$$\begin{aligned}Z &= \widetilde{\mathbf{h}}^H \widetilde{\mathbf{y}} \qquad (5.11)\\ &= \widetilde{\mathbf{h}}^H \widetilde{\mathbf{d}}_0 X + \widetilde{\mathbf{h}}^H \widetilde{\mathbf{v}}\\ &= X_{\text{fd}} + V_{\text{rn}},\end{aligned}$$

where Z is the estimate of the desired signal, X,

$$\begin{aligned} X_{\text{fd}} &= (\mathbf{h}_1 \otimes \mathbf{h}_2)^H (\mathbf{d}_{1,0} \otimes \mathbf{d}_{2,0}) X \\ &= \left(\mathbf{h}_1^H \mathbf{d}_{1,0}\right) \left(\mathbf{h}_2^H \mathbf{d}_{2,0}\right) X \end{aligned} \tag{5.12}$$

is the filtered desired signal, and

$$V_{\text{rn}} = (\mathbf{h}_1 \otimes \mathbf{h}_2)^H \widetilde{\mathbf{v}} \tag{5.13}$$

is the residual noise. We deduce that the variance of Z is

$$\phi_Z = \phi_X \left|\mathbf{h}_1^H \mathbf{d}_{1,0}\right|^2 \left|\mathbf{h}_2^H \mathbf{d}_{2,0}\right|^2 + \phi_{\widetilde{V}_1} (\mathbf{h}_1 \otimes \mathbf{h}_2)^H \mathbf{\Gamma}_{\widetilde{\mathbf{v}}} (\mathbf{h}_1 \otimes \mathbf{h}_2) . \tag{5.14}$$

In our context, the distortionless constraint in the direction of the desired source, i.e., $\theta = 0$, is often required, i.e.,

$$\widetilde{\mathbf{h}}^H \widetilde{\mathbf{d}}_0 = \left(\mathbf{h}_1^H \mathbf{d}_{1,0}\right) \left(\mathbf{h}_2^H \mathbf{d}_{2,0}\right) = 1. \tag{5.15}$$

Therefore, we will always choose $\mathbf{h}_1^H \mathbf{d}_{1,0} = \mathbf{h}_2^H \mathbf{d}_{2,0} = 1$.

It is of course possible to use a complex-valued linear filter, $\mathbf{h}_{\text{C}}$, of length $M_1 M_2$ and apply it to $\widetilde{\mathbf{y}}$ resulting in

$$Z_{\text{C}} = \mathbf{h}_{\text{C}}^H \widetilde{\mathbf{y}}, \tag{5.16}$$

which corresponds to the conventional linear beamforming technique combined with an NULA. This processing, obviously, does not exploit the particular structure of the steering vector $\widetilde{\mathbf{d}}_\theta$. The fundamental difference between $\mathbf{h}_{\text{C}}$ and $\widetilde{\mathbf{h}}$ is that in the former one, $M_1 M_2$ coefficients need to be estimated while in the latter one, only $M_1 + M_2$ coefficients have to be estimated. Also, there is much more flexibility with $\widetilde{\mathbf{h}}$ than with $\mathbf{h}_{\text{C}}$.

5.3 Illustrative Example

Let us take the simplest example of $M_1 = 2$ and $M_2 = 3$. In this case, the steering vectors of the two virtual ULAs are

$$\mathbf{d}_{1,\theta} = \begin{bmatrix} 1 & e^{-\jmath 3\varpi(\theta)} \end{bmatrix}^T, \tag{5.17}$$

$$\mathbf{d}_{2,\theta} = \begin{bmatrix} 1 & e^{-\jmath 2\varpi(\theta)} & e^{-\jmath 4\varpi(\theta)} \end{bmatrix}^T, \tag{5.18}$$

and, as a result, the steering vector of the physical NULA (composed of 6 sensors) is

$$\begin{aligned} \widetilde{\mathbf{d}}_\theta &= \mathbf{d}_{1,\theta} \otimes \mathbf{d}_{2,\theta} \\ &= \begin{bmatrix} 1 & e^{-\jmath 2\varpi(\theta)} & e^{-\jmath 4\varpi(\theta)} & e^{-\jmath 3\varpi(\theta)} & e^{-\jmath 5\varpi(\theta)} & e^{-\jmath 7\varpi(\theta)} \end{bmatrix}^T . \end{aligned} \tag{5.19}$$

We can reorder $\widetilde{\mathbf{d}}_\theta$ as

$$\overline{\mathbf{d}}_\theta = \begin{bmatrix} 1 & e^{-\jmath 2\varpi(\theta)} & e^{-\jmath 3\varpi(\theta)} & e^{-\jmath 4\varpi(\theta)} & e^{-\jmath 5\varpi(\theta)} & e^{-\jmath 7\varpi(\theta)} \end{bmatrix}^T, \tag{5.20}$$

so that now the elements of $\overline{\mathbf{d}}_\theta$ correspond to the sensors of the physical NULA, with minimum and maximum interelement spacings of δ and 2δ, respectively.

The signal model is then

$$\begin{aligned} \widetilde{\mathbf{y}} &= \begin{bmatrix} \widetilde{Y}_1 & \widetilde{Y}_2 & \widetilde{Y}_3 & \widetilde{Y}_4 & \widetilde{Y}_5 & \widetilde{Y}_6 \end{bmatrix}^T \\ &= \widetilde{\mathbf{x}} + \widetilde{\mathbf{v}} \\ &= \widetilde{\mathbf{d}}_0 X + \widetilde{\mathbf{v}}, \end{aligned} \tag{5.21}$$

or, equivalently,

$$\begin{aligned} \overline{\mathbf{y}} &= \begin{bmatrix} \overline{Y}_1 & \overline{Y}_2 & \overline{Y}_3 & \overline{Y}_4 & \overline{Y}_5 & \overline{Y}_6 \end{bmatrix}^T \\ &= \overline{\mathbf{x}} + \overline{\mathbf{v}} \\ &= \overline{\mathbf{d}}_0 X + \overline{\mathbf{v}}, \end{aligned} \tag{5.22}$$

where $\overline{\mathbf{y}}$ and $\overline{\mathbf{v}}$ are ordered with respect to $\overline{\mathbf{d}}_0$.

The Kronecker product filter is

$$\begin{aligned} \widetilde{\mathbf{h}} &= \mathbf{h}_1 \otimes \mathbf{h}_2 \\ &= \begin{bmatrix} H_{1,1}H_{2,1} & H_{1,1}H_{2,2} & H_{1,1}H_{2,3} & H_{1,2}H_{2,1} & H_{1,2}H_{2,2} & H_{1,2}H_{2,3} \end{bmatrix}^T, \end{aligned} \tag{5.23}$$

where H_{1,m_1}, $m_1 = 1, 2$ and H_{2,m_2}, $m_2 = 1, 2, 3$ are the coefficients of the linear filters $\mathbf{h}_1$ and $\mathbf{h}_2$, respectively. We deduce that the ordered filter is

$$\overline{\mathbf{h}} = \begin{bmatrix} H_{1,1}H_{2,1} & H_{1,1}H_{2,2} & H_{1,2}H_{2,1} & H_{1,1}H_{2,3} & H_{1,2}H_{2,2} & H_{1,2}H_{2,3} \end{bmatrix}^T. \tag{5.24}$$

As a consequence, beamforming with Kronecker product filters is

$$\begin{aligned} Z &= \widetilde{\mathbf{h}}^H \widetilde{\mathbf{y}} \\ &= \overline{\mathbf{h}}^H \overline{\mathbf{y}} \\ &= \widetilde{\mathbf{h}}^H \widetilde{\mathbf{d}}_0 X + \widetilde{\mathbf{h}}^H \widetilde{\mathbf{v}} \\ &= \overline{\mathbf{h}}^H \overline{\mathbf{d}}_0 X + \overline{\mathbf{h}}^H \overline{\mathbf{v}} \\ &= X_{\text{fd}} + V_{\text{rn}}. \end{aligned} \tag{5.25}$$

5.4 Performance Measures

The first important measure is the beampattern or directivity pattern, which describes the sensitivity of the beamformer to a plane wave impinging on the NULA from the direction θ. It can be expressed as

$$\begin{aligned}\mathcal{B}_\theta\left(\widetilde{\mathbf{h}}\right) &= \widetilde{\mathbf{d}}_\theta^H \widetilde{\mathbf{h}} \\ &= \left(\mathbf{d}_{1,\theta}^H \mathbf{h}_1\right)\left(\mathbf{d}_{2,\theta}^H \mathbf{h}_2\right) \\ &= \mathcal{B}_{1,\theta}\left(\mathbf{h}_1\right) \times \mathcal{B}_{2,\theta}\left(\mathbf{h}_2\right),\end{aligned} \tag{5.26}$$

where

$$\begin{aligned}\mathcal{B}_{1,\theta}\left(\mathbf{h}_1\right) &= \mathbf{d}_{1,\theta}^H \mathbf{h}_1 \\ &= \sum_{m_1=1}^{M_1} H_{1,m_1} e^{\jmath(m_1-1)M_2\varpi(\theta)}\end{aligned} \tag{5.27}$$

and

$$\begin{aligned}\mathcal{B}_{2,\theta}\left(\mathbf{h}_2\right) &= \mathbf{d}_{2,\theta}^H \mathbf{h}_2 \\ &= \sum_{m_2=1}^{M_2} H_{2,m_2} e^{\jmath(m_2-1)M_1\varpi(\theta)},\end{aligned} \tag{5.28}$$

with H_{1,m_1}, $m_1 = 1, 2, \ldots, M_1$ and H_{2,m_2}, $m_2 = 1, 2, \ldots, M_2$ being the coefficients of the filters $\mathbf{h}_1$ and $\mathbf{h}_2$, respectively. Let $\mathcal{Z}_1 = e^{\jmath M_2\varpi(\theta)}$ and $\mathcal{Z}_2 = e^{\jmath M_1\varpi(\theta)}$, we can express the global beampattern as a polynomial in two variables, which is the product of two polynomials (of degrees $M_1 - 1$ and $M_2 - 1$) in one variable each, i.e.,

$$\begin{aligned}\mathcal{B}\left(\mathcal{Z}_1, \mathcal{Z}_2\right) &= \mathcal{B}_1\left(\mathcal{Z}_1\right) \times \mathcal{B}_2\left(\mathcal{Z}_2\right) \\ &= \left(\sum_{m_1=1}^{M_1} H_{1,m_1} \mathcal{Z}_1^{m_1-1}\right)\left(\sum_{m_2=1}^{M_2} H_{2,m_2} \mathcal{Z}_2^{m_2-1}\right).\end{aligned} \tag{5.29}$$

From this perspective, we can see that this beampattern has at most $M_1 + M_2 - 2$ distinct nulls (between 0 and π), while the beampattern of a conventional linear array with $M_1 M_2$ sensors has at most $M_1 M_2 - 1$ distinct nulls (between 0 and π). The fact that $\mathcal{B}_\theta\left(\widetilde{\mathbf{h}}\right)$ can be expressed as the product of two beampatterns is an interesting property that can be exploited in the design of flexible beamformers.

Given that the first sensor is the reference, we can define the input SNR with respect to this reference as

$$\text{iSNR} = \frac{\phi_X}{\phi_{\widetilde{V}_1}}. \tag{5.30}$$

The output SNR is defined [from the variance of Z, see (5.14)] as

$$\begin{aligned}
\text{oSNR}\left(\widetilde{\mathbf{h}}\right) &= \phi_X \frac{\left|\widetilde{\mathbf{h}}^H \widetilde{\mathbf{d}}_0\right|^2}{\widetilde{\mathbf{h}}^H \mathbf{\Phi}_{\widetilde{\mathbf{v}}} \widetilde{\mathbf{h}}} \\
&= \frac{\phi_X}{\phi_{\widetilde{V}_1}} \times \frac{\left|\widetilde{\mathbf{h}}^H \widetilde{\mathbf{d}}_0\right|^2}{\widetilde{\mathbf{h}}^H \mathbf{\Gamma}_{\widetilde{\mathbf{v}}} \widetilde{\mathbf{h}}}.
\end{aligned} \tag{5.31}$$

The definition of the gain in SNR is easily derived from the previous definitions, i.e.,

$$\begin{aligned}
\mathcal{G}\left(\widetilde{\mathbf{h}}\right) &= \frac{\text{oSNR}\left(\widetilde{\mathbf{h}}\right)}{\text{iSNR}} \\
&= \frac{\left|\widetilde{\mathbf{h}}^H \widetilde{\mathbf{d}}_0\right|^2}{\widetilde{\mathbf{h}}^H \mathbf{\Gamma}_{\widetilde{\mathbf{v}}} \widetilde{\mathbf{h}}}.
\end{aligned} \tag{5.32}$$

The most convenient way to evaluate the sensitivity of the NULA to some of its imperfections is via the WNG, which is defined by taking $\mathbf{\Gamma}_{\widetilde{\mathbf{v}}} = \mathbf{I}_{M_1 M_2}$ in (5.32), where $\mathbf{I}_{M_1 M_2}$ is the $M_1 M_2 \times M_1 M_2$ identity matrix, i.e.,

$$\begin{aligned}
\mathcal{W}\left(\widetilde{\mathbf{h}}\right) &= \frac{\left|\widetilde{\mathbf{h}}^H \widetilde{\mathbf{d}}_0\right|^2}{\widetilde{\mathbf{h}}^H \widetilde{\mathbf{h}}} \\
&= \frac{\left|\mathbf{h}_1^H \mathbf{d}_{1,0}\right|^2}{\mathbf{h}_1^H \mathbf{h}_1} \times \frac{\left|\mathbf{h}_2^H \mathbf{d}_{2,0}\right|^2}{\mathbf{h}_2^H \mathbf{h}_2} \\
&= \mathcal{W}_1\left(\mathbf{h}_1\right) \times \mathcal{W}_2\left(\mathbf{h}_2\right),
\end{aligned} \tag{5.33}$$

where

$$\mathcal{W}_1\left(\mathbf{h}_1\right) = \frac{\left|\mathbf{h}_1^H \mathbf{d}_{1,0}\right|^2}{\mathbf{h}_1^H \mathbf{h}_1} \tag{5.34}$$

and

$$\mathcal{W}_2\left(\mathbf{h}_2\right) = \frac{\left|\mathbf{h}_2^H \mathbf{d}_{2,0}\right|^2}{\mathbf{h}_2^H \mathbf{h}_2}. \tag{5.35}$$

Obviously, the WNG of the NULA is simply the product of the WNGs of the two virtual ULAs described in Section 5.1. It is easy to check that

$$\mathcal{W}\left(\widetilde{\mathbf{h}}\right) \leq M_1 M_2, \ \forall \widetilde{\mathbf{h}}. \tag{5.36}$$

Another important measure, which quantifies how the microphone array performs in the presence of presence of spatial noise and reverberation is the DF:

$$\begin{aligned}
\mathcal{D}\left(\widetilde{\mathbf{h}}\right) &= \frac{\left|\mathcal{B}_0\left(\widetilde{\mathbf{h}}\right)\right|^2}{\frac{1}{2}\int_0^{\pi}\left|\mathcal{B}_\theta\left(\widetilde{\mathbf{h}}\right)\right|^2 \sin\theta d\theta} \\
&= \frac{\left|\mathcal{B}_{1,0}\left(\mathbf{h}_1\right)\right|^2\left|\mathcal{B}_{2,0}\left(\mathbf{h}_2\right)\right|^2}{\frac{1}{2}\int_0^{\pi}\left|\mathcal{B}_{1,\theta}\left(\mathbf{h}_1\right)\right|^2\left|\mathcal{B}_{2,\theta}\left(\mathbf{h}_2\right)\right|^2 \sin\theta d\theta} \\
&= \frac{\left|\widetilde{\mathbf{h}}^H\widetilde{\mathbf{d}}_0\right|^2}{\widetilde{\mathbf{h}}^H\widetilde{\mathbf{\Gamma}}\widetilde{\mathbf{h}}},
\end{aligned} \tag{5.37}$$

where $\widetilde{\mathbf{\Gamma}}$ is defined in (5.9). One can verify that

$$\mathcal{D}\left(\widetilde{\mathbf{h}}\right) \leq \widetilde{\mathbf{d}}_0^H\widetilde{\mathbf{\Gamma}}^{-1}\widetilde{\mathbf{d}}_0, \ \forall \widetilde{\mathbf{h}}. \tag{5.38}$$

We observe that contrary to the beampattern and the WNG, the DF of the NULA cannot be factorized, i.e.,

$$\mathcal{D}\left(\widetilde{\mathbf{h}}\right) \neq \mathcal{D}_1\left(\mathbf{h}_1\right) \times \mathcal{D}_2\left(\mathbf{h}_2\right), \tag{5.39}$$

where

$$\begin{aligned}
\mathcal{D}_1\left(\mathbf{h}_1\right) &= \frac{\left|\mathcal{B}_{1,0}\left(\mathbf{h}_1\right)\right|^2}{\frac{1}{2}\int_0^{\pi}\left|\mathcal{B}_{1,\theta}\left(\mathbf{h}_1\right)\right|^2 \sin\theta d\theta} \\
&= \frac{\left|\mathbf{h}_1^H\mathbf{d}_{1,0}\right|^2}{\mathbf{h}_1^H\mathbf{\Gamma}_1\mathbf{h}_1},
\end{aligned} \tag{5.40}$$

$$\begin{aligned}
\mathcal{D}_2\left(\mathbf{h}_2\right) &= \frac{\left|\mathcal{B}_{2,0}\left(\mathbf{h}_2\right)\right|^2}{\frac{1}{2}\int_0^{\pi}\left|\mathcal{B}_{2,\theta}\left(\mathbf{h}_2\right)\right|^2 \sin\theta d\theta} \\
&= \frac{\left|\mathbf{h}_2^H\mathbf{d}_{2,0}\right|^2}{\mathbf{h}_2^H\mathbf{\Gamma}_2\mathbf{h}_2},
\end{aligned} \tag{5.41}$$

with

$$\mathbf{\Gamma}_1 = \frac{1}{2}\int_0^{\pi} \mathbf{d}_{1,\theta}\mathbf{d}_{1,\theta}^H \sin\theta d\theta, \tag{5.42}$$

$$\mathbf{\Gamma}_2 = \frac{1}{2}\int_0^{\pi} \mathbf{d}_{2,\theta}\mathbf{d}_{2,\theta}^H \sin\theta d\theta. \tag{5.43}$$

From the mean value theorem, we have

$$\begin{aligned}
&\int_0^{\pi} \left|\mathcal{B}_{1,\theta}\left(\mathbf{h}_1\right)\right|^2 \left|\mathcal{B}_{2,\theta}\left(\mathbf{h}_2\right)\right|^2 \sin\theta d\theta \\
&= \left|\mathcal{B}_{1,\theta_1}\left(\mathbf{h}_1\right)\right|^2 \int_0^{\pi} \left|\mathcal{B}_{2,\theta}\left(\mathbf{h}_2\right)\right|^2 \sin\theta d\theta \\
&= \left|\mathcal{B}_{2,\theta_2}\left(\mathbf{h}_2\right)\right|^2 \int_0^{\pi} \left|\mathcal{B}_{1,\theta}\left(\mathbf{h}_1\right)\right|^2 \sin\theta d\theta,
\end{aligned} \tag{5.44}$$

where $\theta_1, \theta_2 \in [0, \pi]$. As a result, we can write the DF in (5.37) as

$$\begin{aligned}
\mathcal{D}\left(\widetilde{\mathbf{h}}\right) &= \frac{\left|\mathcal{B}_{1,0}\left(\mathbf{h}_1\right)\right|^2}{\left|\mathcal{B}_{1,\theta_1}\left(\mathbf{h}_1\right)\right|^2} \times \mathcal{D}_2\left(\mathbf{h}_2\right) \\
&= \frac{\left|\mathcal{B}_{2,0}\left(\mathbf{h}_2\right)\right|^2}{\left|\mathcal{B}_{2,\theta_2}\left(\mathbf{h}_2\right)\right|^2} \times \mathcal{D}_1\left(\mathbf{h}_1\right) \\
&= \frac{\left|\mathcal{B}_{1,0}\left(\mathbf{h}_1\right)\right|}{\left|\mathcal{B}_{1,\theta_1}\left(\mathbf{h}_1\right)\right|} \times \frac{\left|\mathcal{B}_{2,0}\left(\mathbf{h}_2\right)\right|}{\left|\mathcal{B}_{2,\theta_2}\left(\mathbf{h}_2\right)\right|} \times \sqrt{\mathcal{D}_1\left(\mathbf{h}_1\right) \times \mathcal{D}_2\left(\mathbf{h}_2\right)}.
\end{aligned} \tag{5.45}$$

Since the filters $\mathbf{h}_1$ and $\mathbf{h}_2$ are always designed in such a way that $\left|\mathcal{B}_{1,0}\left(\mathbf{h}_1\right)\right| \geq \left|\mathcal{B}_{1,\theta_1}\left(\mathbf{h}_1\right)\right|$ and $\left|\mathcal{B}_{2,0}\left(\mathbf{h}_2\right)\right| \geq \left|\mathcal{B}_{2,\theta_2}\left(\mathbf{h}_2\right)\right|$, we deduce that

$$\mathcal{D}\left(\widetilde{\mathbf{h}}\right) \geq \sqrt{\mathcal{D}_1\left(\mathbf{h}_1\right) \times \mathcal{D}_2\left(\mathbf{h}_2\right)}. \tag{5.46}$$

It can be verified that

$$\mathbf{h}_1 \otimes \mathbf{h}_2 = \left(\mathbf{h}_1 \otimes \mathbf{I}_{M_2}\right)\mathbf{h}_2 \tag{5.47}$$

$$= \left(\mathbf{I}_{M_1} \otimes \mathbf{h}_2\right)\mathbf{h}_1, \tag{5.48}$$

where $\mathbf{I}_{M_1}$ and $\mathbf{I}_{M_2}$ are the identity matrices of sizes $M_1 \times M_1$ and $M_2 \times M_2$, respectively. When $\mathbf{h}_2$ is fixed and given, using the distortionless constraint, i.e., $\mathbf{h}_2^H \mathbf{d}_{2,0} = 1$, and thanks to (5.48), we can write the DF as

$$\begin{aligned}
\mathcal{D}\left(\mathbf{h}_1 | \mathbf{h}_2\right) &= \frac{\left|\mathcal{B}_{1,0}\left(\mathbf{h}_1\right)\right|^2}{\mathbf{h}_1^H \mathbf{\Gamma}_{\mathbf{h}_2} \mathbf{h}_1} \\
&= \frac{\left|\mathbf{h}_1^H \mathbf{d}_{1,0}\right|^2}{\mathbf{h}_1^H \mathbf{\Gamma}_{\mathbf{h}_2} \mathbf{h}_1},
\end{aligned} \tag{5.49}$$

where

$$\begin{aligned}\mathbf{\Gamma}_{\mathbf{h}_2} &= \frac{1}{2}\int_0^\pi \mathbf{d}_{1,\theta}\mathbf{d}_{1,\theta}^H \left|\mathcal{B}_{2,\theta}\left(\mathbf{h}_2\right)\right|^2 \sin\theta d\theta \\ &= \left(\mathbf{I}_{M_1}\otimes\mathbf{h}_2\right)^H \widetilde{\mathbf{\Gamma}}\left(\mathbf{I}_{M_1}\otimes\mathbf{h}_2\right).\end{aligned} \tag{5.50}$$

In the same way, when $\mathbf{h}_1$ is fixed and given, using the distortionless constraint, i.e., $\mathbf{h}_1^H\mathbf{d}_{1,0} = 1$, and thanks to (5.47), we can express the DF as

$$\begin{aligned}\mathcal{D}\left(\mathbf{h}_2|\mathbf{h}_1\right) &= \frac{\left|\mathcal{B}_{2,0}\left(\mathbf{h}_2\right)\right|^2}{\mathbf{h}_2^H\mathbf{\Gamma}_{\mathbf{h}_1}\mathbf{h}_2} \\ &= \frac{\left|\mathbf{h}_2^H\mathbf{d}_{2,0}\right|^2}{\mathbf{h}_2^H\mathbf{\Gamma}_{\mathbf{h}_1}\mathbf{h}_2},\end{aligned} \tag{5.51}$$

where

$$\begin{aligned}\mathbf{\Gamma}_{\mathbf{h}_1} &= \frac{1}{2}\int_0^\pi \mathbf{d}_{2,\theta}\mathbf{d}_{2,\theta}^H \left|\mathcal{B}_{1,\theta}\left(\mathbf{h}_1\right)\right|^2 \sin\theta d\theta \\ &= \left(\mathbf{h}_1\otimes\mathbf{I}_{M_2}\right)^H \widetilde{\mathbf{\Gamma}}\left(\mathbf{h}_1\otimes\mathbf{I}_{M_2}\right).\end{aligned} \tag{5.52}$$

Finally, the last measure of interest in this chapter is the FBR, which is mathematically defined as [10]

$$\begin{aligned}\mathcal{F}\left(\widetilde{\mathbf{h}}\right) &= \frac{\int_0^{\pi/2}\left|\mathcal{B}_\theta\left(\widetilde{\mathbf{h}}\right)\right|^2 \sin\theta d\theta}{\int_{\pi/2}^{\pi}\left|\mathcal{B}_\theta\left(\widetilde{\mathbf{h}}\right)\right|^2 \sin\theta d\theta} \\ &= \frac{\widetilde{\mathbf{h}}^H\widetilde{\mathbf{\Gamma}}_{\mathrm{f}}\widetilde{\mathbf{h}}}{\widetilde{\mathbf{h}}^H\widetilde{\mathbf{\Gamma}}_{\mathrm{b}}\widetilde{\mathbf{h}}},\end{aligned} \tag{5.53}$$

where

$$\widetilde{\mathbf{\Gamma}}_{\mathrm{f}} = \int_0^{\pi/2}\widetilde{\mathbf{d}}_\theta\widetilde{\mathbf{d}}_\theta^H \sin\theta d\theta, \tag{5.54}$$

$$\widetilde{\mathbf{\Gamma}}_{\mathrm{b}} = \int_{\pi/2}^{\pi}\widetilde{\mathbf{d}}_\theta\widetilde{\mathbf{d}}_\theta^H \sin\theta d\theta. \tag{5.55}$$

The FBR of the NULA cannot be factorized, i.e.,

$$\mathcal{F}\left(\widetilde{\mathbf{h}}\right) \neq \mathcal{F}_1\left(\mathbf{h}_1\right)\times\mathcal{F}_2\left(\mathbf{h}_2\right), \tag{5.56}$$

where

$$\mathcal{F}_1(\mathbf{h}_1) = \frac{\displaystyle\int_0^{\pi/2} |\mathcal{B}_{1,\theta}(\mathbf{h}_1)|^2 \sin\theta d\theta}{\displaystyle\int_{\pi/2}^{\pi} |\mathcal{B}_{1,\theta}(\mathbf{h}_1)|^2 \sin\theta d\theta} = \frac{\mathbf{h}_1^H \mathbf{\Gamma}_{\mathrm{f},1}\mathbf{h}_1}{\mathbf{h}_1^H \mathbf{\Gamma}_{\mathrm{b},1}\mathbf{h}_1}, \tag{5.57}$$

$$\mathcal{F}_2(\mathbf{h}_2) = \frac{\displaystyle\int_0^{\pi/2} |\mathcal{B}_{2,\theta}(\mathbf{h}_2)|^2 \sin\theta d\theta}{\displaystyle\int_{\pi/2}^{\pi} |\mathcal{B}_{2,\theta}(\mathbf{h}_2)|^2 \sin\theta d\theta} = \frac{\mathbf{h}_2^H \mathbf{\Gamma}_{\mathrm{f},2}\mathbf{h}_2}{\mathbf{h}_2^H \mathbf{\Gamma}_{\mathrm{b},2}\mathbf{h}_2}, \tag{5.58}$$

with

$$\mathbf{\Gamma}_{\mathrm{f},1} = \int_0^{\pi/2} \mathbf{d}_{1,\theta}\mathbf{d}_{1,\theta}^H \sin\theta d\theta, \tag{5.59}$$

$$\mathbf{\Gamma}_{\mathrm{b},1} = \int_{\pi/2}^{\pi} \mathbf{d}_{1,\theta}\mathbf{d}_{1,\theta}^H \sin\theta d\theta, \tag{5.60}$$

$$\mathbf{\Gamma}_{\mathrm{f},2} = \int_0^{\pi/2} \mathbf{d}_{2,\theta}\mathbf{d}_{2,\theta}^H \sin\theta d\theta, \tag{5.61}$$

$$\mathbf{\Gamma}_{\mathrm{b},2} = \int_{\pi/2}^{\pi} \mathbf{d}_{2,\theta}\mathbf{d}_{2,\theta}^H \sin\theta d\theta. \tag{5.62}$$

When $\mathbf{h}_2$ is fixed and given, and thanks to (5.48), we can write the FBR as

$$\mathcal{F}(\mathbf{h}_1|\mathbf{h}_2) = \frac{\mathbf{h}_1^H \mathbf{\Gamma}_{\mathrm{f},\mathbf{h}_2}\mathbf{h}_1}{\mathbf{h}_1^H \mathbf{\Gamma}_{\mathrm{b},\mathbf{h}_2}\mathbf{h}_1}, \tag{5.63}$$

where

$$\mathbf{\Gamma}_{\mathrm{f},\mathbf{h}_2} = \int_0^{\pi/2} \mathbf{d}_{1,\theta}\mathbf{d}_{1,\theta}^H |\mathcal{B}_{2,\theta}(\mathbf{h}_2)|^2 \sin\theta d\theta = (\mathbf{I}_{M_1} \otimes \mathbf{h}_2)^H \widetilde{\mathbf{\Gamma}}_{\mathrm{f}} (\mathbf{I}_{M_1} \otimes \mathbf{h}_2), \tag{5.64}$$

$$\mathbf{\Gamma}_{\mathrm{b},\mathbf{h}_2} = \int_{\pi/2}^{\pi} \mathbf{d}_{1,\theta}\mathbf{d}_{1,\theta}^H |\mathcal{B}_{2,\theta}(\mathbf{h}_2)|^2 \sin\theta d\theta = (\mathbf{I}_{M_1} \otimes \mathbf{h}_2)^H \widetilde{\mathbf{\Gamma}}_{\mathrm{b}} (\mathbf{I}_{M_1} \otimes \mathbf{h}_2). \tag{5.65}$$

In the same way, when $\mathbf{h}_1$ is fixed and given, and thanks to (5.47), we can write the FBR as

$$\mathcal{F}(\mathbf{h}_2|\mathbf{h}_1) = \frac{\mathbf{h}_2^H \mathbf{\Gamma}_{\mathrm{f},\mathbf{h}_1} \mathbf{h}_2}{\mathbf{h}_2^H \mathbf{\Gamma}_{\mathrm{b},\mathbf{h}_1} \mathbf{h}_2}, \tag{5.66}$$

where

$$\begin{aligned} \mathbf{\Gamma}_{\mathrm{f},\mathbf{h}_1} &= \int_0^{\pi/2} \mathbf{d}_{2,\theta} \mathbf{d}_{2,\theta}^H \left| \mathcal{B}_{1,\theta}(\mathbf{h}_1) \right|^2 \sin\theta d\theta \qquad (5.67) \\ &= (\mathbf{h}_1 \otimes \mathbf{I}_{M_2})^H \widetilde{\mathbf{\Gamma}}_{\mathrm{f}} (\mathbf{h}_1 \otimes \mathbf{I}_{M_2}), \\ \mathbf{\Gamma}_{\mathrm{b},\mathbf{h}_1} &= \int_{\pi/2}^{\pi} \mathbf{d}_{2,\theta} \mathbf{d}_{2,\theta}^H \left| \mathcal{B}_{1,\theta}(\mathbf{h}_1) \right|^2 \sin\theta d\theta \qquad (5.68) \\ &= (\mathbf{h}_1 \otimes \mathbf{I}_{M_2})^H \widetilde{\mathbf{\Gamma}}_{\mathrm{b}} (\mathbf{h}_1 \otimes \mathbf{I}_{M_2}). \end{aligned}$$

5.5 Examples of Beamformers

In this section, we derive some useful and optimal examples of Kronecker product beamformers. Of course, many more can be deduced depending on the applications at hand.

5.5.1 Delay and Sum

Given the structure of the WNG of $\widetilde{\mathbf{h}}$, it is clear that the maximization of this gain is equivalent to maximizing $\mathcal{W}_1(\mathbf{h}_1)$ and $\mathcal{W}_2(\mathbf{h}_2)$ separately. Taking into account the distortionless constraints, we easily get the DS beamformers at the two virtual ULAs:

$$\mathbf{h}_{1,\mathrm{DS}} = \frac{\mathbf{d}_{1,0}}{M_1}, \tag{5.69}$$

$$\mathbf{h}_{2,\mathrm{DS}} = \frac{\mathbf{d}_{2,0}}{M_2}. \tag{5.70}$$

As a consequence, the DS beamformer corresponding to the physical NULA is

$$\begin{aligned} \widetilde{\mathbf{h}}_{\mathrm{DS}} &= \mathbf{h}_{1,\mathrm{DS}} \otimes \mathbf{h}_{2,\mathrm{DS}} \\ &= \frac{\mathbf{d}_{1,0} \otimes \mathbf{d}_{2,0}}{M_1 M_2} \\ &= \frac{\widetilde{\mathbf{d}}_0}{M_1 M_2}. \end{aligned} \tag{5.71}$$

It is clear that

$$\mathcal{W}\left(\widetilde{\mathbf{h}}_{\mathrm{DS}}\right) = M_1 M_2, \tag{5.72}$$

which is the same with the one obtained with a ULA composed of $M_1 M_2$ sensors. However, the beampattern of the DS beamformer is

$$\begin{aligned}
\mathcal{B}_\theta\left(\widetilde{\mathbf{h}}_{\mathrm{DS}}\right) &= \mathcal{B}_{1,\theta}\left(\mathbf{h}_{1,\mathrm{DS}}\right) \times \mathcal{B}_{2,\theta}\left(\mathbf{h}_{2,\mathrm{DS}}\right) \\
&= \frac{1}{M_1 M_2}\left(\mathbf{d}_{1,\theta}^H \mathbf{d}_{1,0}\right)\left(\mathbf{d}_{2,\theta}^H \mathbf{d}_{2,0}\right) \\
&= \frac{\left\{1 - e^{\jmath M_1 M_2 [\varpi(\theta) - \varpi(0)]}\right\}^2}{M_1 M_2 \left\{1 - e^{\jmath M_1 [\varpi(\theta) - \varpi(0)]}\right\}\left\{1 - e^{\jmath M_2 [\varpi(\theta) - \varpi(0)]}\right\}},
\end{aligned} \tag{5.73}$$

which is very much different from the one obtained with a ULA composed of $M_1 M_2$ sensors.

Figure 5.1 displays the directivity patterns of the DS beamformer, $\widetilde{\mathbf{h}}_{\mathrm{DS}}$, for $f = 2$ kHz, $\delta = 1$ cm, $M_1 = 2$, and different numbers of sensors M_2. Figure 5.2 shows plots of the DFs and WNGs of the DS beamformer as a function of frequency for $\delta = 1$ cm, $M_1 = 2$, and different numbers of sensors M_2. We observe that as the number of sensors increases, the width of the main beam and the level of side lobes decrease. As the number of sensors increases, both the DF and the WNG of the DS beamformer increase.

5.5.2 Partial Superdirective

Obviously, there are different ways to derive some partial superdirective (or partial hypercardioid) beamformers thanks to the definition of the DF of $\widetilde{\mathbf{h}}$.

In the first approach, we assume that $\mathbf{h}_2$ is fixed. We may take $\mathbf{h}_2 = \mathbf{h}_{2,\mathrm{DS}}$ for the second virtual ULA. Substituting this filter into (5.49), we get

$$\mathcal{D}\left(\mathbf{h}_1 | \mathbf{h}_{2,\mathrm{DS}}\right) = \frac{\left|\mathbf{h}_1^H \mathbf{d}_{1,0}\right|^2}{\mathbf{h}_1^H \mathbf{\Gamma}_{\mathbf{h}_{2,\mathrm{DS}}} \mathbf{h}_1}, \tag{5.74}$$

where

$$\begin{aligned}
\mathbf{\Gamma}_{\mathbf{h}_{2,\mathrm{DS}}} &= \frac{1}{2}\int_0^\pi \mathbf{d}_{1,\theta}\mathbf{d}_{1,\theta}^H \left|\mathcal{B}_{2,\theta}\left(\mathbf{h}_{2,\mathrm{DS}}\right)\right|^2 \sin\theta d\theta \\
&= \left(\mathbf{I}_{M_1} \otimes \mathbf{h}_{2,\mathrm{DS}}\right)^H \widetilde{\mathbf{\Gamma}} \left(\mathbf{I}_{M_1} \otimes \mathbf{h}_{2,\mathrm{DS}}\right).
\end{aligned} \tag{5.75}$$

The maximization of $\mathcal{D}\left(\mathbf{h}_1 | \mathbf{h}_{2,\mathrm{DS}}\right)$ gives the superdirective beamformer at the first virtual ULA:

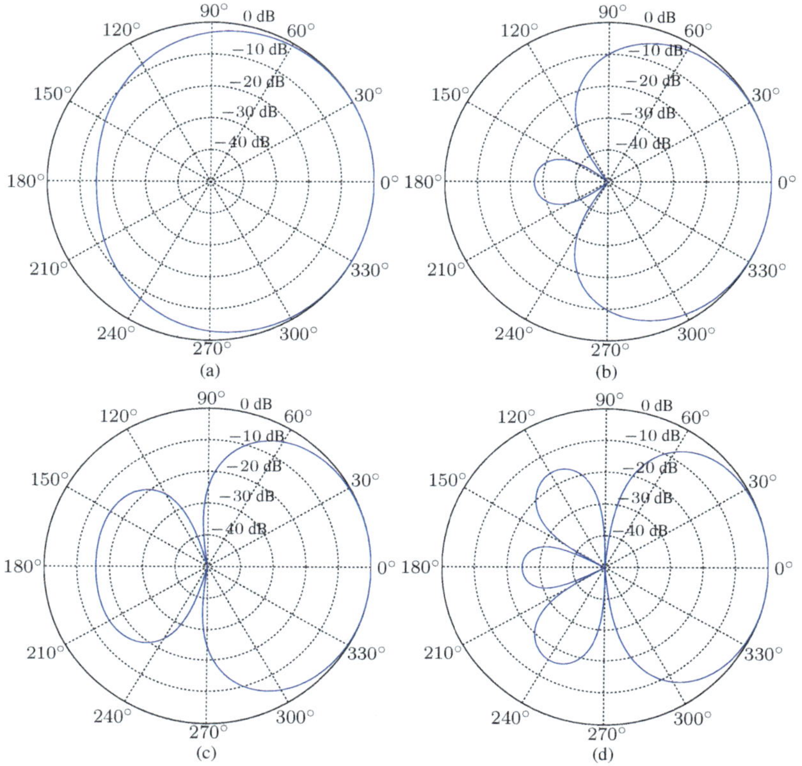

Fig. 5.1 Beampatterns of the DS beamformer, $\widetilde{\mathbf{h}}_{\mathrm{DS}}$, for $f = 2$ kHz, $\delta = 1$ cm, $M_1 = 2$, and different numbers of sensors M_2: (a) $M_2 = 3$, (b) $M_2 = 5$, (c) $M_2 = 7$, and (d) $M_2 = 9$.

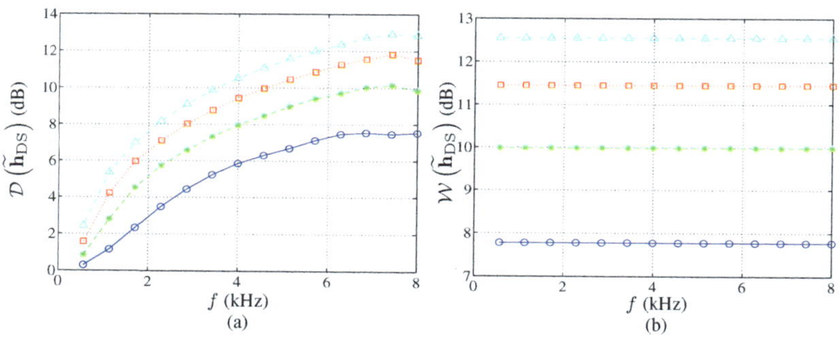

Fig. 5.2 Performance of the DS beamformer, $\widetilde{\mathbf{h}}_{\mathrm{DS}}$, as a function of frequency for $\delta = 1$ cm, $M_1 = 2$, and different numbers of sensors M_2: $M_2 = 3$ (solid line with circles), $M_2 = 5$ (dashed line with asterisks), $M_2 = 7$ (dotted line with squares), and $M_2 = 9$ (dash-dot line with triangles). (a) DF and (b) WNG.

$$\mathbf{h}_{1,\mathrm{S1}} = \frac{\mathbf{\Gamma}_{\mathbf{h}_{2,\mathrm{DS}}}^{-1} \mathbf{d}_{1,0}}{\mathbf{d}_{1,0}^H \mathbf{\Gamma}_{\mathbf{h}_{2,\mathrm{DS}}}^{-1} \mathbf{d}_{1,0}}. \tag{5.76}$$

Therefore, our first partial superdirective (PS) beamformer is

$$\widetilde{\mathbf{h}}_{\mathrm{PS1}} = \mathbf{h}_{1,\mathrm{S1}} \otimes \mathbf{h}_{2,\mathrm{DS}}. \tag{5.77}$$

We deduce that the WNG and the beampattern are, respectively,

$$\mathcal{W}\left(\widetilde{\mathbf{h}}_{\mathrm{PS1}}\right) = M_2 \mathcal{W}\left(\mathbf{h}_{1,\mathrm{S1}}\right) \tag{5.78}$$

and

$$\mathcal{B}_{\theta}\left(\widetilde{\mathbf{h}}_{\mathrm{PS1}}\right) = \mathcal{B}_{1,\theta}\left(\mathbf{h}_{1,\mathrm{S1}}\right) \times \mathcal{B}_{2,\theta}\left(\mathbf{h}_{2,\mathrm{DS}}\right). \tag{5.79}$$

Figure 5.3 displays the directivity patterns of the first partial superdirective beamformer, $\widetilde{\mathbf{h}}_{\mathrm{PS1}}$, for $f = 2$ kHz, $\delta = 1$ cm, $M_1 = 2$, and different numbers of sensors M_2. Figure 5.4 shows plots of the DFs and WNGs of the first partial superdirective beamformer as a function of frequency for $\delta = 1$ cm, $M_1 = 2$, and different numbers of sensors M_2. We observe that as the number of sensors increases, the width of the main beam decreases, and both the DF and the WNG of the first partial superdirective beamformer increase. Compared to the DS beamformer, the first partial superdirective beamformer yields higher DF, but lower WNG (compare Figs 5.2 and 5.4).

In the second approach, we assume that $\mathbf{h}_1$ is fixed, i.e., $\mathbf{h}_1 = \mathbf{h}_{1,\mathrm{DS}}$ for the first virtual ULA. Substituting this filter into (5.51), we get

$$\mathcal{D}\left(\mathbf{h}_2 | \mathbf{h}_{1,\mathrm{DS}}\right) = \frac{\left|\mathbf{h}_2^H \mathbf{d}_{2,0}\right|^2}{\mathbf{h}_2^H \mathbf{\Gamma}_{\mathbf{h}_{1,\mathrm{DS}}} \mathbf{h}_2}, \tag{5.80}$$

where

$$\begin{aligned} \mathbf{\Gamma}_{\mathbf{h}_{1,\mathrm{DS}}} &= \frac{1}{2} \int_0^{\pi} \mathbf{d}_{2,\theta} \mathbf{d}_{2,\theta}^H \left|\mathcal{B}_{1,\theta}\left(\mathbf{h}_{1,\mathrm{DS}}\right)\right|^2 \sin\theta d\theta \\ &= \left(\mathbf{h}_{1,\mathrm{DS}} \otimes \mathbf{I}_{M_2}\right)^H \widetilde{\mathbf{\Gamma}} \left(\mathbf{h}_{1,\mathrm{DS}} \otimes \mathbf{I}_{M_2}\right). \end{aligned} \tag{5.81}$$

The maximization of $\mathcal{D}\left(\mathbf{h}_2 | \mathbf{h}_{1,\mathrm{DS}}\right)$ gives the superdirective beamformer at the second virtual ULA:

$$\mathbf{h}_{2,\mathrm{S2}} = \frac{\mathbf{\Gamma}_{\mathbf{h}_{1,\mathrm{DS}}}^{-1} \mathbf{d}_{2,0}}{\mathbf{d}_{2,0}^H \mathbf{\Gamma}_{\mathbf{h}_{1,\mathrm{DS}}}^{-1} \mathbf{d}_{2,0}}. \tag{5.82}$$

As a result, our second partial superdirective beamformer is

$$\widetilde{\mathbf{h}}_{\mathrm{PS2}} = \mathbf{h}_{1,\mathrm{DS}} \otimes \mathbf{h}_{2,\mathrm{S2}}. \tag{5.83}$$

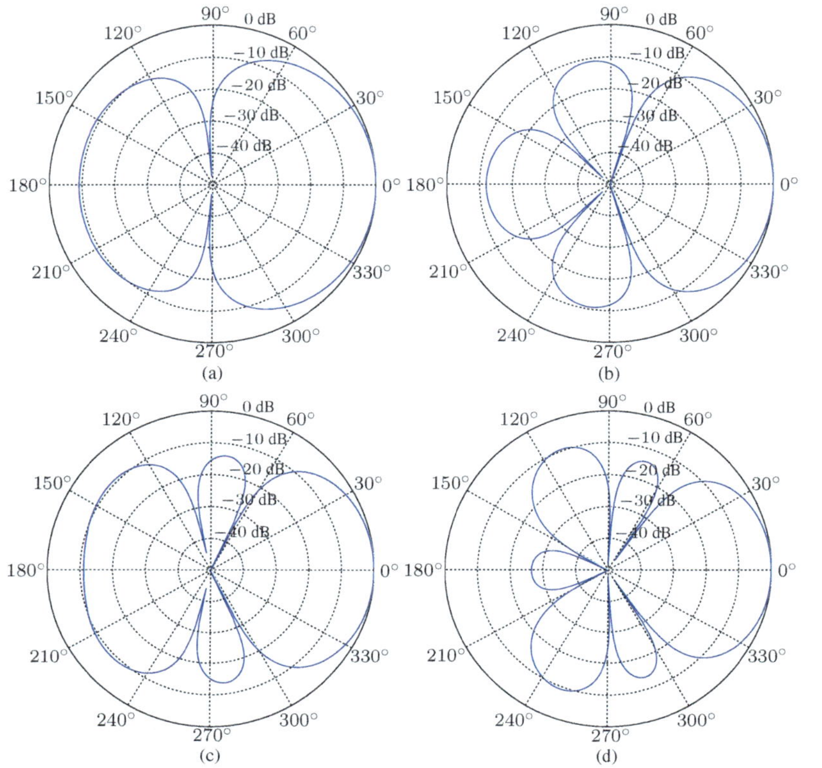

Fig. 5.3 Beampatterns of the first partial superdirective beamformer, $\widetilde{\mathbf{h}}_{\mathrm{PS1}}$, for $f = 2$ kHz, $\delta = 1$ cm, $M_1 = 2$, and different numbers of sensors M_2: (a) $M_2 = 3$, (b) $M_2 = 5$, (c) $M_2 = 7$, and (d) $M_2 = 9$.

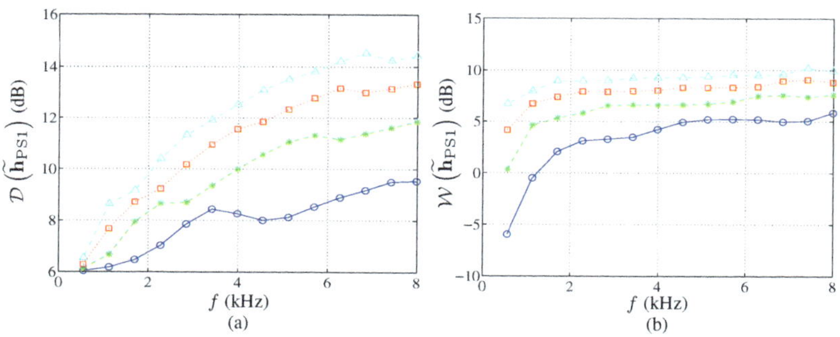

Fig. 5.4 Performance of the first partial superdirective beamformer, $\widetilde{\mathbf{h}}_{\mathrm{PS1}}$, as a function of frequency for $\delta = 1$ cm, $M_1 = 2$, and different numbers of sensors M_2: $M_2 = 3$ (solid line with circles), $M_2 = 5$ (dashed line with asterisks), $M_2 = 7$ (dotted line with squares), and $M_2 = 9$ (dash-dot line with triangles). (a) DF and (b) WNG.

We deduce that the WNG and the beampattern are, respectively,

$$\mathcal{W}\left(\widetilde{\mathbf{h}}_{\text{PS2}}\right) = M_1 \mathcal{W}\left(\mathbf{h}_{2,\text{S2}}\right) \tag{5.84}$$

and

$$\mathcal{B}_\theta\left(\widetilde{\mathbf{h}}_{\text{PS2}}\right) = \mathcal{B}_{1,\theta}\left(\mathbf{h}_{1,\text{DS}}\right) \times \mathcal{B}_{2,\theta}\left(\mathbf{h}_{2,\text{S2}}\right). \tag{5.85}$$

From the two superdirective beamformers derived above for the virtual ULAs, we find the third approach:

$$\widetilde{\mathbf{h}}_{\text{PS3}} = \mathbf{h}_{1,\text{S1}} \otimes \mathbf{h}_{2,\text{S2}}. \tag{5.86}$$

Finally, the last approach is obtained by maximizing separately the two DFs, $\mathcal{D}_1\left(\mathbf{h}_1\right)$ and $\mathcal{D}_2\left(\mathbf{h}_2\right)$, of the virtual ULAs. We get

$$\mathbf{h}_{1,\text{S}} = \frac{\boldsymbol{\Gamma}_1^{-1}\mathbf{d}_{1,0}}{\mathbf{d}_{1,0}^H\boldsymbol{\Gamma}_1^{-1}\mathbf{d}_{1,0}}, \tag{5.87}$$

$$\mathbf{h}_{2,\text{S}} = \frac{\boldsymbol{\Gamma}_2^{-1}\mathbf{d}_{2,0}}{\mathbf{d}_{2,0}^H\boldsymbol{\Gamma}_2^{-1}\mathbf{d}_{2,0}}. \tag{5.88}$$

As a result, the partial superdirective beamformer of the fourth approach is simply the Kronecker product of the two above filters, i.e.,

$$\widetilde{\mathbf{h}}_{\text{PS4}} = \mathbf{h}_{1,\text{S}} \otimes \mathbf{h}_{2,\text{S}}. \tag{5.89}$$

Figure 5.5 displays the directivity patterns of the fourth partial superdirective beamformer, $\widetilde{\mathbf{h}}_{\text{PS4}}$, for $f = 2$ kHz, $\delta = 1$ cm, $M_1 = 2$, and different numbers of sensors M_2. Figure 5.6 shows plots of the DFs and WNGs of the fourth partial superdirective beamformer as a function of frequency for $\delta = 1$ cm, $M_1 = 2$, and different numbers of sensors M_2. We observe that as the number of sensors increases, the width of the main beam decreases, the DF increases, but the WNG decreases. Compared to the first partial superdirective beamformer, the fourth partial superdirective beamformer yields higher DF, but much lower WNG, especially at low frequencies (compare Figs 5.4 and 5.6).

5.5.3 Superdirective

The superdirective beamformer is obtained by maximizing $\mathcal{D}\left(\widetilde{\mathbf{h}}\right)$ in (5.37). But this DF cannot be maximized directly and an iterative algorithm is needed.

We start by taking

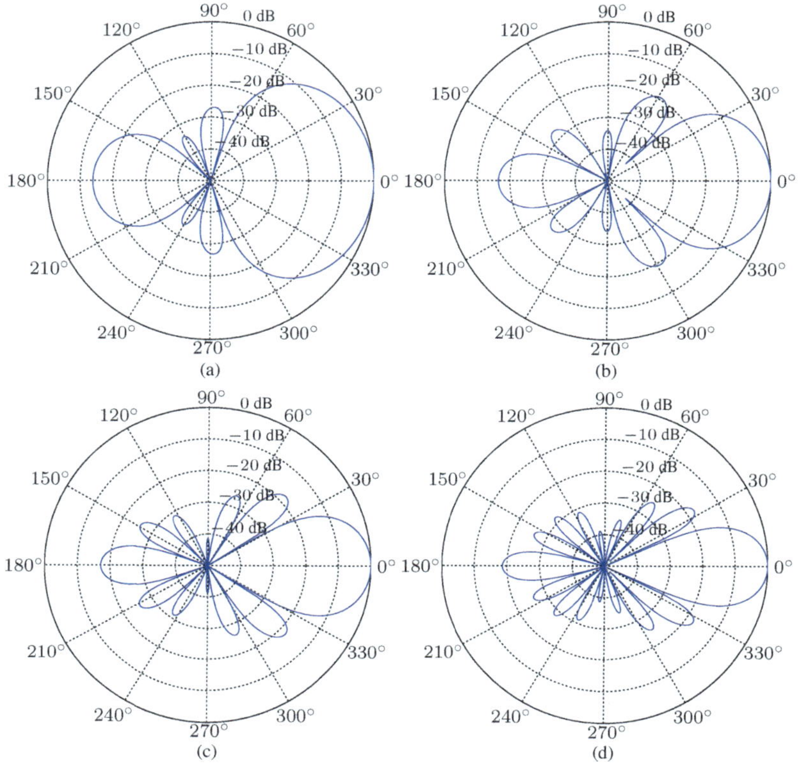

Fig. 5.5 Beampatterns of the fourth partial superdirective beamformer, $\widetilde{\mathbf{h}}_{\mathrm{PS4}}$, for $f = 2$ kHz, $\delta = 1$ cm, $M_1 = 2$, and different numbers of sensors M_2: (a) $M_2 = 3$, (b) $M_2 = 5$, (c) $M_2 = 7$, and (d) $M_2 = 9$.

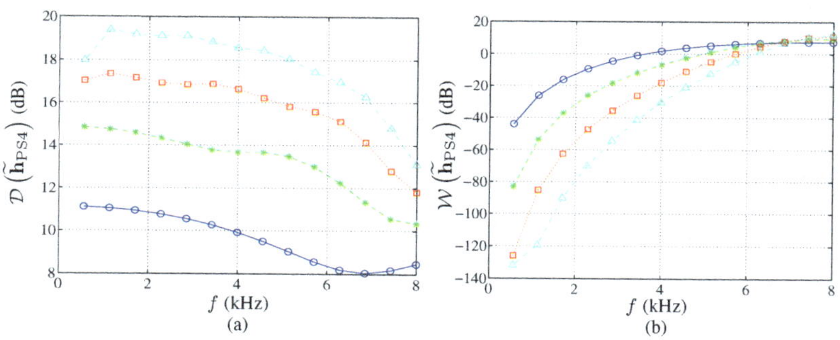

Fig. 5.6 Performance of the fourth partial superdirective beamformer, $\widetilde{\mathbf{h}}_{\mathrm{PS4}}$, as a function of frequency for $\delta = 1$ cm, $M_1 = 2$, and different numbers of sensors M_2: $M_2 = 3$ (solid line with circles), $M_2 = 5$ (dashed line with asterisks), $M_2 = 7$ (dotted line with squares), and $M_2 = 9$ (dash-dot line with triangles). (a) DF and (b) WNG.

$$\begin{aligned}\mathbf{h}_2^{(0)} &= \mathbf{h}_{2,\mathrm{S}} \\ &= \frac{\mathbf{\Gamma}_2^{-1}\mathbf{d}_{2,0}}{\mathbf{d}_{2,0}^H\mathbf{\Gamma}_2^{-1}\mathbf{d}_{2,0}}.\end{aligned} \tag{5.90}$$

Substituting $\mathbf{h}_2^{(0)}$ into (5.50), we get

$$\mathbf{\Gamma}_{\mathbf{h}_2^{(0)}} = \left(\mathbf{I}_{M_1} \otimes \mathbf{h}_2^{(0)}\right)^H \widetilde{\mathbf{\Gamma}} \left(\mathbf{I}_{M_1} \otimes \mathbf{h}_2^{(0)}\right). \tag{5.91}$$

Now, substituting this expression into the DF in (5.49), we obtain at iteration 1:

$$\mathcal{D}\left(\mathbf{h}_1^{(1)}|\mathbf{h}_2^{(0)}\right) = \frac{\left|\left(\mathbf{h}_1^{(1)}\right)^H \mathbf{d}_{1,0}\right|^2}{\left(\mathbf{h}_1^{(1)}\right)^H \mathbf{\Gamma}_{\mathbf{h}_2^{(0)}}\mathbf{h}_1^{(1)}}. \tag{5.92}$$

The maximization of $\mathcal{D}\left(\mathbf{h}_1^{(1)}|\mathbf{h}_2^{(0)}\right)$ with respect to $\mathbf{h}_1^{(1)}$ gives

$$\mathbf{h}_1^{(1)} = \frac{\mathbf{\Gamma}_{\mathbf{h}_2^{(0)}}^{-1}\mathbf{d}_{1,0}}{\mathbf{d}_{1,0}^H\mathbf{\Gamma}_{\mathbf{h}_2^{(0)}}^{-1}\mathbf{d}_{1,0}}. \tag{5.93}$$

Using $\mathbf{h}_1^{(1)}$ in (5.52), we get

$$\mathbf{\Gamma}_{\mathbf{h}_1^{(1)}} = \left(\mathbf{h}_1^{(1)} \otimes \mathbf{I}_{M_2}\right)^H \widetilde{\mathbf{\Gamma}} \left(\mathbf{h}_1^{(1)} \otimes \mathbf{I}_{M_2}\right). \tag{5.94}$$

As a result, the DF in (5.51) is

$$\mathcal{D}\left(\mathbf{h}_2^{(1)}|\mathbf{h}_1^{(1)}\right) = \frac{\left|\left(\mathbf{h}_2^{(1)}\right)^H \mathbf{d}_{2,0}\right|^2}{\left(\mathbf{h}_2^{(1)}\right)^H \mathbf{\Gamma}_{\mathbf{h}_1^{(1)}}\mathbf{h}_2^{(1)}}, \tag{5.95}$$

whose maximization with respect to $\mathbf{h}_2^{(1)}$ gives

$$\mathbf{h}_2^{(1)} = \frac{\mathbf{\Gamma}_{\mathbf{h}_1^{(1)}}^{-1}\mathbf{d}_{2,0}}{\mathbf{d}_{2,0}^H\mathbf{\Gamma}_{\mathbf{h}_1^{(1)}}^{-1}\mathbf{d}_{2,0}}. \tag{5.96}$$

Continuing the iterations up to the iteration n, we easily get for the first filter:

$$\mathbf{h}_1^{(n)} = \frac{\mathbf{\Gamma}_{\mathbf{h}_2^{(n-1)}}^{-1} \mathbf{d}_{1,0}}{\mathbf{d}_{1,0}^H \mathbf{\Gamma}_{\mathbf{h}_2^{(n-1)}}^{-1} \mathbf{d}_{1,0}}, \tag{5.97}$$

with

$$\mathbf{\Gamma}_{\mathbf{h}_2^{(n-1)}} = \left(\mathbf{I}_{M_1} \otimes \mathbf{h}_2^{(n-1)}\right)^H \widetilde{\mathbf{\Gamma}} \left(\mathbf{I}_{M_1} \otimes \mathbf{h}_2^{(n-1)}\right), \tag{5.98}$$

and for the second filter:

$$\mathbf{h}_2^{(n)} = \frac{\mathbf{\Gamma}_{\mathbf{h}_1^{(n)}}^{-1} \mathbf{d}_{2,0}}{\mathbf{d}_{2,0}^H \mathbf{\Gamma}_{\mathbf{h}_1^{(n)}}^{-1} \mathbf{d}_{2,0}}, \tag{5.99}$$

with

$$\mathbf{\Gamma}_{\mathbf{h}_1^{(n)}} = \left(\mathbf{h}_1^{(n)} \otimes \mathbf{I}_{M_2}\right)^H \widetilde{\mathbf{\Gamma}} \left(\mathbf{h}_1^{(n)} \otimes \mathbf{I}_{M_2}\right). \tag{5.100}$$

Finally, we deduce that the superdirective beamformer is at iteration n:

$$\widetilde{\mathbf{h}}_{\mathrm{SD}}^{(n)} = \mathbf{h}_1^{(n)} \otimes \mathbf{h}_2^{(n)}. \tag{5.101}$$

Figure 5.7 displays the directivity patterns of the superdirective beamformer, $\widetilde{\mathbf{h}}_{\mathrm{SD}}^{(n)}$, for $f = 2$ kHz, $\delta = 5$ mm, $M_1 = 3$, and $M_2 = 4$, obtained at the iteration n for several values of n. Figure 5.8 shows plots of the DFs and WNGs of the superdirective beamformer as a function of frequency for $\delta = 5$ mm, $M_1 = 3$, $M_2 = 4$, and several values of n. We observe that the DF of the superdirective beamformer increases at each iteration, and roughly converges after three iterations, while the WNG decreases at each iteration. Compared with the partial superdirective beamformers, the superdirective beamformer yields higher DF, but lower WNG.

5.5.4 Dipole

Let us show how to design the first-order dipole with the NULA. Fundamentally, the first virtual ULA will shape the desired pattern while the second virtual ULA will help reduce white noise amplification. Therefore, we take $M_1 = 2$ and $M_2 \geq 3$.

The first-order dipole has a null at the angle $\pi/2$. Taking into account the distortionless constraint, we get the linear system of equations:

$$\mathbf{D}_{1,\pi/2}\mathbf{h}_1 = \begin{bmatrix} 1 \\ 0 \end{bmatrix}, \tag{5.102}$$

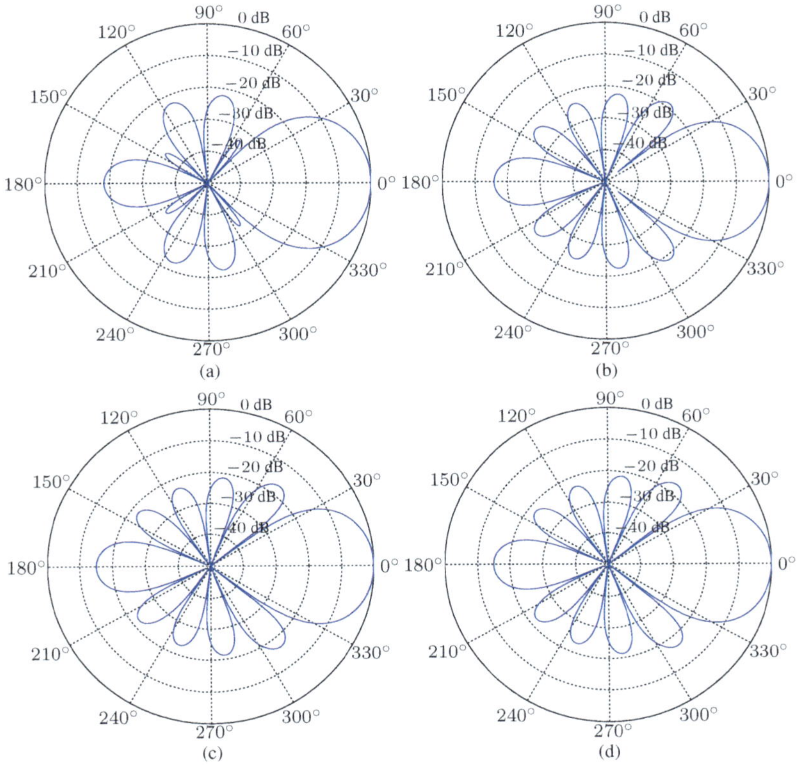

Fig. 5.7 Beampatterns of the superdirective beamformer, $\widetilde{\mathbf{h}}_{\mathrm{SD}}^{(n)}$, for $f = 2$ kHz, $\delta = 5$ mm, $M_1 = 3$, and $M_2 = 4$, obtained at the iteration n: (a) $n = 1$, (b) $n = 2$, (c) $n = 3$, and (d) $n = 10$.

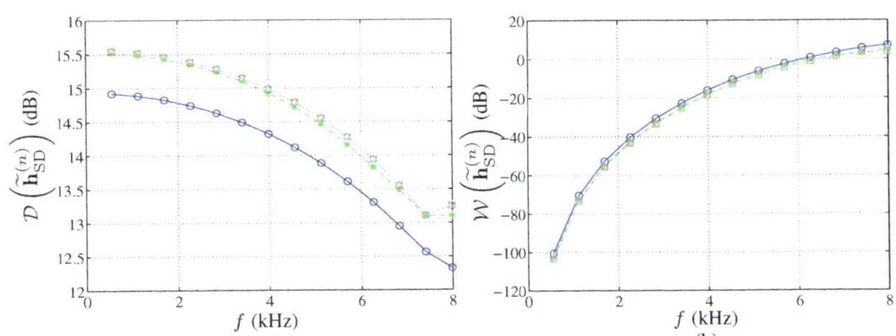

Fig. 5.8 Performance of the superdirective beamformer, $\widetilde{\mathbf{h}}_{\mathrm{SD}}^{(n)}$, as a function of frequency for $\delta = 5$ mm, $M_1 = 3$, $M_2 = 4$, and several values of n: $n = 1$ (solid line with circles), $n = 2$ (dashed line with asterisks), $n = 3$ (dotted line with squares), and $n = 10$ (dash-dot line with triangles). (a) DF and (b) WNG.

where

$$\mathbf{D}_{1,\pi/2} = \begin{bmatrix} \mathbf{d}_{1,0}^H \\ \mathbf{d}_{1,\pi/2}^H \end{bmatrix} \tag{5.103}$$

is a 2×2 matrix. As a result,

$$\mathbf{h}_1 = \mathbf{D}_{1,\pi/2}^{-1} \begin{bmatrix} 1 \\ 0 \end{bmatrix}. \tag{5.104}$$

With $\mathbf{h}_1$, the beampattern of the NULA will always have a null at $\pi/2$, no matter how $\mathbf{h}_2$ is designed. Another characteristic of the dipole is that it has a one at the angle π. Therefore, this constraint should be included in the design of $\mathbf{h}_2$. Given that we want to maximize the WNG with the second virtual ULA, our criterion to optimize is

$$\min_{\mathbf{h}_2} \mathbf{h}_2^H \mathbf{h}_2 \quad \text{subject to} \quad \mathbf{C}_{2,\pi} \mathbf{h}_2 = \begin{bmatrix} 1 \\ 1 \end{bmatrix}, \tag{5.105}$$

where

$$\mathbf{C}_{2,\pi} = \begin{bmatrix} \mathbf{d}_{2,0}^H \\ \mathbf{d}_{2,\pi}^H \end{bmatrix} \tag{5.106}$$

is a $2 \times M_2$ matrix. We find that

$$\mathbf{h}_{2,\mathrm{D1}} = \mathbf{C}_{2,\pi}^H \left(\mathbf{C}_{2,\pi} \mathbf{C}_{2,\pi}^H \right)^{-1} \begin{bmatrix} 1 \\ 1 \end{bmatrix}. \tag{5.107}$$

From (5.104) and (5.107), we deduce that the designed filter for the first-order dipole with the NULA is

$$\widetilde{\mathbf{h}}_{\mathrm{D1}} = \mathbf{h}_1 \otimes \mathbf{h}_{2,\mathrm{D1}}. \tag{5.108}$$

Figure 5.9 displays the directivity patterns of the first-order dipole with the NULA, $\widetilde{\mathbf{h}}_{\mathrm{D1}}$, for $f = 2$ kHz, $\delta = 5$ mm, $M_1 = 2$, and different numbers of sensors M_2. Figure 5.10 shows plots of the DFs and WNGs of the first-order dipole with the NULA as a function of frequency for $\delta = 5$ mm, $M_1 = 2$, and different numbers of sensors M_2. We observe that as the number of sensors increases, the WNG of the first-order dipole at low frequencies increases.

If we want to increase the order of the dipole, we can still take $\mathbf{h}_1$ as it is in (5.104) but add a null constraint at $\pi/2$ to $\mathbf{h}_2$. In this case, the criterion to optimize is

$$\min_{\mathbf{h}_2} \mathbf{h}_2^H \mathbf{h}_2 \quad \text{subject to} \quad \mathbf{C}_{2,\pi/2,\pi} \mathbf{h}_2 = \begin{bmatrix} 1 \\ 0 \\ 1 \end{bmatrix}, \tag{5.109}$$

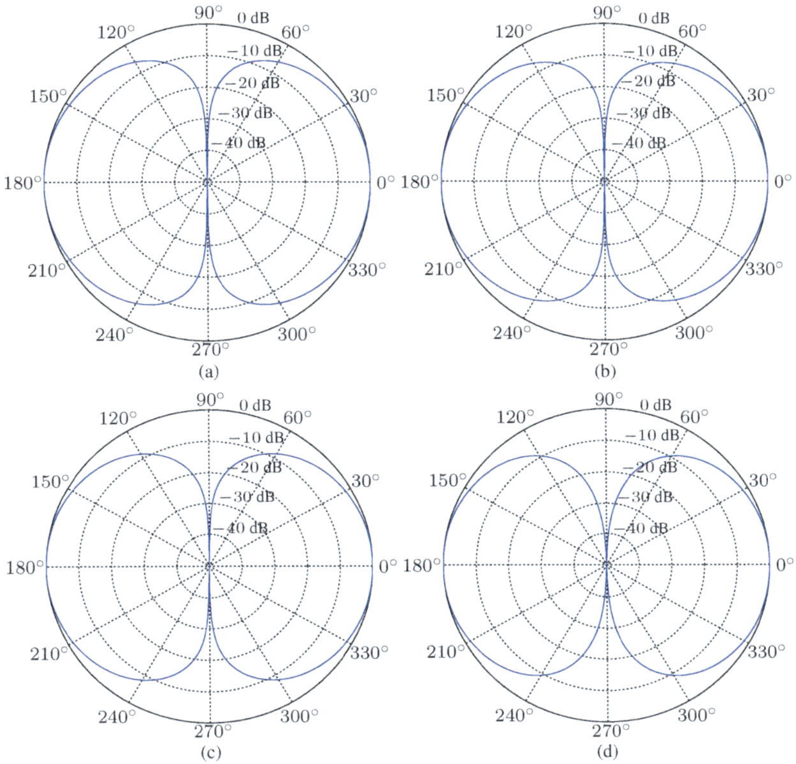

Fig. 5.9 Beampatterns of the first-order dipole with the NULA, $\widetilde{\mathbf{h}}_{\mathrm{D1}}$, for $f = 2$ kHz, $\delta = 5$ mm, $M_1 = 2$, and different numbers of sensors M_2: (a) $M_2 = 3$, (b) $M_2 = 5$, (c) $M_2 = 7$, and (d) $M_2 = 9$.

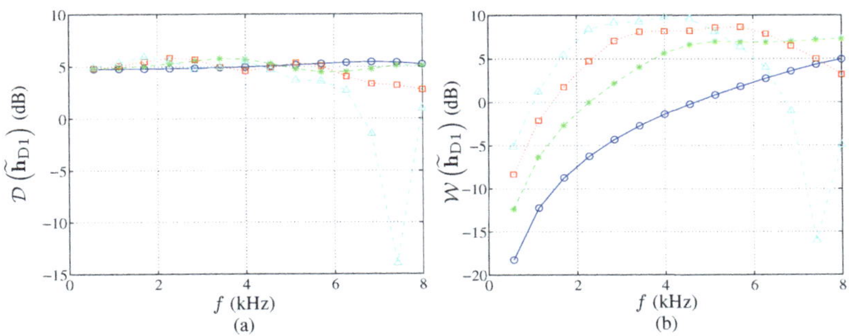

Fig. 5.10 Performance of the first-order dipole with the NULA, $\widetilde{\mathbf{h}}_{\mathrm{D1}}$, as a function of frequency for $\delta = 5$ mm, $M_1 = 2$, and different numbers of sensors M_2: $M_2 = 3$ (solid line with circles), $M_2 = 5$ (dashed line with asterisks), $M_2 = 7$ (dotted line with squares), and $M_2 = 9$ (dash-dot line with triangles). (a) DF and (b) WNG.

where

$$\mathbf{C}_{2,\pi/2,\pi} = \begin{bmatrix} \mathbf{d}_{2,0}^H \\ \mathbf{d}_{2,\pi/2}^H \\ \mathbf{d}_{2,\pi}^H \end{bmatrix} \quad (5.110)$$

is a $3 \times M_2$ matrix. We find that

$$\mathbf{h}_{2,\mathrm{D2}} = \mathbf{C}_{2,\pi/2,\pi}^H \left(\mathbf{C}_{2,\pi/2,\pi} \mathbf{C}_{2,\pi/2\pi}^H \right)^{-1} \begin{bmatrix} 1 \\ 0 \\ 1 \end{bmatrix}. \quad (5.111)$$

Therefore, a higher-order dipole with the NULA is

$$\widetilde{\mathbf{h}}_{\mathrm{D2}} = \mathbf{h}_1 \otimes \mathbf{h}_{2,\mathrm{D2}}. \quad (5.112)$$

Figure 5.11 displays the directivity patterns of the higher-order dipole with the NULA, $\widetilde{\mathbf{h}}_{\mathrm{D2}}$, for $f = 2$ kHz, $\delta = 5$ mm, $M_1 = 2$, and different numbers of sensors M_2. Figure 5.12 shows plots of the DFs and WNGs of $\widetilde{\mathbf{h}}_{\mathrm{D2}}$ as a function of frequency for $\delta = 5$ mm, $M_1 = 2$, and different numbers of sensors M_2. We observe that as the number of sensors increases, at low frequencies the WNG of the higher-order dipole increases, but the DF decreases. Compared to the first-order dipole with the NULA, the higher-order dipole yields higher DF, but lower WNG at low frequencies (compare Figs 5.10 and 5.12).

5.5.5 Supercardioid

The supercardioid is obtained by maximizing the FBR. To fully maximize the FBR in (5.53), we need to derive an iterative algorithm.

At iteration 0, we may take

$$\mathbf{h}_2^{(0)} = \frac{\mathbf{t}_2}{\mathbf{d}_{2,0}^H \mathbf{t}_2}, \quad (5.113)$$

where $\mathbf{t}_2$ is the eigenvector corresponding to the maximum eigenvalue of the matrix $\mathbf{\Gamma}_{\mathrm{b},2}^{-1}\mathbf{\Gamma}_{\mathrm{f},2}$ (see Section 5.4). Substituting $\mathbf{h}_2^{(0)}$ into (5.64) and (5.65), we get

$$\mathbf{\Gamma}_{\mathrm{f},\mathbf{h}_2^{(0)}} = \left(\mathbf{I}_{M_1} \otimes \mathbf{h}_2^{(0)} \right)^H \widetilde{\mathbf{\Gamma}}_{\mathrm{f}} \left(\mathbf{I}_{M_1} \otimes \mathbf{h}_2^{(0)} \right), \quad (5.114)$$

$$\mathbf{\Gamma}_{\mathrm{b},\mathbf{h}_2^{(0)}} = \left(\mathbf{I}_{M_1} \otimes \mathbf{h}_2^{(0)} \right)^H \widetilde{\mathbf{\Gamma}}_{\mathrm{b}} \left(\mathbf{I}_{M_1} \otimes \mathbf{h}_2^{(0)} \right). \quad (5.115)$$

Now, plugging these expressions into the FBR in (5.63), we obtain at iteration 1:

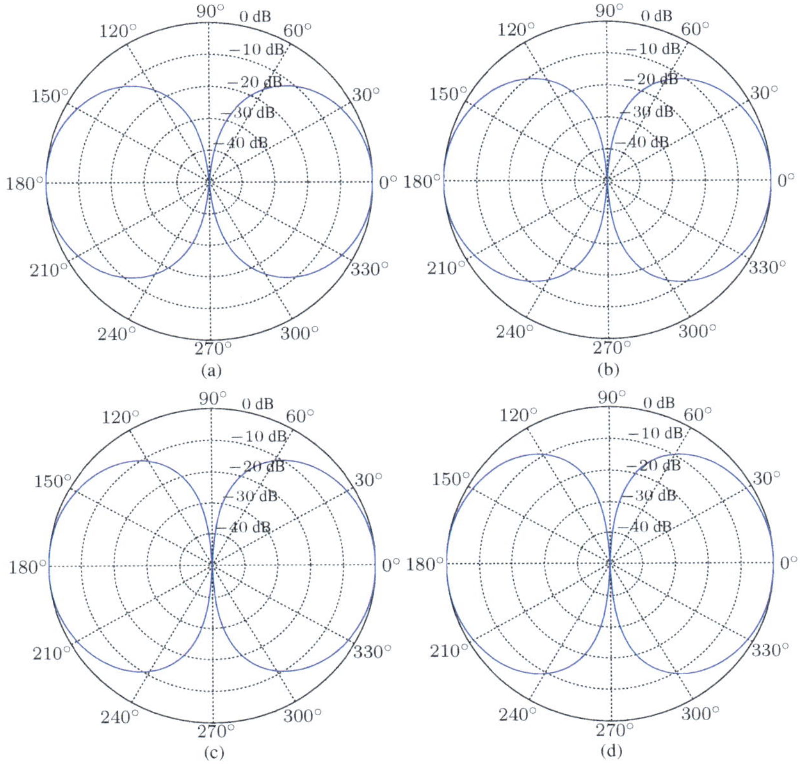

Fig. 5.11 Beampatterns of a higher-order dipole with the NULA, $\widetilde{\mathbf{h}}_{\mathrm{D}2}$, for $f = 2$ kHz, $\delta = 5$ mm, $M_1 = 2$, and different numbers of sensors M_2: (a) $M_2 = 3$, (b) $M_2 = 5$, (c) $M_2 = 7$, and (d) $M_2 = 9$.

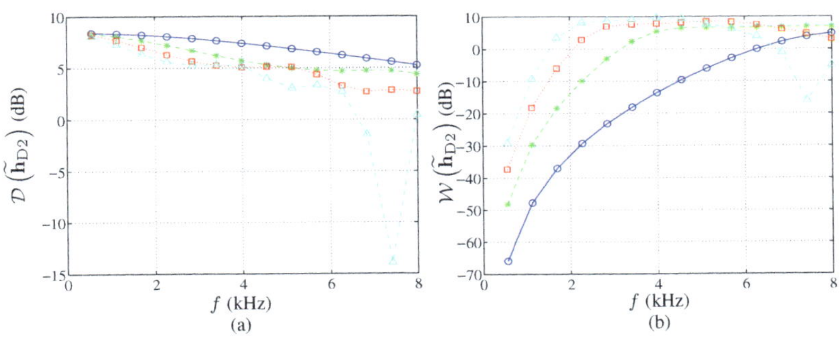

Fig. 5.12 Performance of a higher-order dipole with the NULA, $\widetilde{\mathbf{h}}_{\mathrm{D}2}$, as a function of frequency for $\delta = 5$ mm, $M_1 = 2$, and different numbers of sensors M_2: $M_2 = 3$ (solid line with circles), $M_2 = 5$ (dashed line with asterisks), $M_2 = 7$ (dotted line with squares), and $M_2 = 9$ (dash-dot line with triangles). (a) DF and (b) WNG.

$$\mathcal{F}\left(\mathbf{h}_1^{(1)}|\mathbf{h}_2^{(0)}\right) = \frac{\left(\mathbf{h}_1^{(1)}\right)^H \mathbf{\Gamma}_{\mathrm{f},\mathbf{h}_2^{(0)}} \mathbf{h}_1^{(1)}}{\left(\mathbf{h}_1^{(1)}\right)^H \mathbf{\Gamma}_{\mathrm{b},\mathbf{h}_2^{(0)}} \mathbf{h}_1^{(1)}}. \tag{5.116}$$

The maximization of $\mathcal{F}\left(\mathbf{h}_1^{(1)}|\mathbf{h}_2^{(0)}\right)$ with respect of $\mathbf{h}_1^{(1)}$ leads to

$$\mathbf{h}_1^{(1)} = \frac{\mathbf{t}_1^{(0)}}{\mathbf{d}_{1,0}^H \mathbf{t}_1^{(0)}}, \tag{5.117}$$

where $\mathbf{t}_1^{(0)}$ is the eigenvector corresponding to the maximum eigenvalue of the matrix $\mathbf{\Gamma}_{\mathrm{b},\mathbf{h}_2^{(0)}}^{-1} \mathbf{\Gamma}_{\mathrm{f},\mathbf{h}_2^{(0)}}$. Using $\mathbf{h}_1^{(1)}$ in (5.67) and (5.68), we get

$$\mathbf{\Gamma}_{\mathrm{f},\mathbf{h}_1^{(1)}} = \left(\mathbf{h}_1^{(1)} \otimes \mathbf{I}_{M_2}\right)^H \widetilde{\mathbf{\Gamma}}_{\mathrm{f}} \left(\mathbf{h}_1^{(1)} \otimes \mathbf{I}_{M_2}\right), \tag{5.118}$$

$$\mathbf{\Gamma}_{\mathrm{b},\mathbf{h}_1^{(1)}} = \left(\mathbf{h}_1^{(1)} \otimes \mathbf{I}_{M_2}\right)^H \widetilde{\mathbf{\Gamma}}_{\mathrm{b}} \left(\mathbf{h}_1^{(1)} \otimes \mathbf{I}_{M_2}\right). \tag{5.119}$$

As a result, the FBR in (5.66) is

$$\mathcal{F}\left(\mathbf{h}_2^{(1)}|\mathbf{h}_1^{(1)}\right) = \frac{\left(\mathbf{h}_2^{(1)}\right)^H \mathbf{\Gamma}_{\mathrm{f},\mathbf{h}_1^{(1)}} \mathbf{h}_2^{(1)}}{\left(\mathbf{h}_2^{(1)}\right)^H \mathbf{\Gamma}_{\mathrm{b},\mathbf{h}_1^{(1)}} \mathbf{h}_2^{(1)}}, \tag{5.120}$$

whose maximization with respect to $\mathbf{h}_2^{(1)}$ gives

$$\mathbf{h}_2^{(1)} = \frac{\mathbf{t}_2^{(1)}}{\mathbf{d}_{2,0}^H \mathbf{t}_2^{(1)}}, \tag{5.121}$$

where $\mathbf{t}_2^{(1)}$ is the eigenvector corresponding to the maximum eigenvalue of the matrix $\mathbf{\Gamma}_{\mathrm{b},\mathbf{h}_1^{(1)}}^{-1} \mathbf{\Gamma}_{\mathrm{f},\mathbf{h}_1^{(1)}}$.

Continuing to iterate up to iteration n, we easily get for the first filter:

$$\mathbf{h}_1^{(n)} = \frac{\mathbf{t}_1^{(n-1)}}{\mathbf{d}_{1,0}^H \mathbf{t}_1^{(n-1)}}, \tag{5.122}$$

where $\mathbf{t}_1^{(n-1)}$ is the eigenvector corresponding to the maximum eigenvalue of the matrix $\mathbf{\Gamma}_{\mathrm{b},\mathbf{h}_2^{(n-1)}}^{-1} \mathbf{\Gamma}_{\mathrm{f},\mathbf{h}_2^{(n-1)}}$, with

$$\boldsymbol{\Gamma}_{\mathrm{f},\mathbf{h}_2^{(n-1)}} = \left(\mathbf{I}_{M_1} \otimes \mathbf{h}_2^{(n-1)}\right)^H \widetilde{\boldsymbol{\Gamma}}_{\mathrm{f}} \left(\mathbf{I}_{M_1} \otimes \mathbf{h}_2^{(n-1)}\right), \tag{5.123}$$

$$\boldsymbol{\Gamma}_{\mathrm{b},\mathbf{h}_2^{(n-1)}} = \left(\mathbf{I}_{M_1} \otimes \mathbf{h}_2^{(n-1)}\right)^H \widetilde{\boldsymbol{\Gamma}}_{\mathrm{b}} \left(\mathbf{I}_{M_1} \otimes \mathbf{h}_2^{(n-1)}\right), \tag{5.124}$$

and for the second filter:

$$\mathbf{h}_2^{(n)} = \frac{\mathbf{t}_2^{(n)}}{\mathbf{d}_{2,0}^H \mathbf{t}_2^{(n)}}, \tag{5.125}$$

where $\mathbf{t}_2^{(n)}$ is the eigenvector corresponding to the maximum eigenvalue of the matrix $\boldsymbol{\Gamma}^{-1}_{\mathrm{b},\mathbf{h}_1^{(n)}} \boldsymbol{\Gamma}_{\mathrm{f},\mathbf{h}_1^{(n)}}$, with

$$\boldsymbol{\Gamma}_{\mathrm{f},\mathbf{h}_1^{(n)}} = \left(\mathbf{h}_1^{(n)} \otimes \mathbf{I}_{M_2}\right)^H \widetilde{\boldsymbol{\Gamma}}_{\mathrm{f}} \left(\mathbf{h}_1^{(n)} \otimes \mathbf{I}_{M_2}\right), \tag{5.126}$$

$$\boldsymbol{\Gamma}_{\mathrm{b},\mathbf{h}_1^{(n)}} = \left(\mathbf{h}_1^{(n)} \otimes \mathbf{I}_{M_2}\right)^H \widetilde{\boldsymbol{\Gamma}}_{\mathrm{b}} \left(\mathbf{h}_1^{(n)} \otimes \mathbf{I}_{M_2}\right). \tag{5.127}$$

Finally, we deduce that the supercardioid beamformer is at iteration n:

$$\widetilde{\mathbf{h}}_{\mathrm{S}}^{(n)} = \mathbf{h}_1^{(n)} \otimes \mathbf{h}_2^{(n)}. \tag{5.128}$$

Figure 5.13 displays the directivity patterns of the supercardioid beamformer, $\widetilde{\mathbf{h}}_{\mathrm{S}}^{(n)}$, for $f = 2$ kHz, $\delta = 5$ mm, $M_1 = 3$, and $M_2 = 4$, obtained at the iteration n for several values of n. Figure 5.14 shows plots of the DFs, the WNGs, and the FBRs of the supercardioid beamformer as a function of frequency for $\delta = 5$ mm, $M_1 = 3$, $M_2 = 4$, and several values of n. We observe that the FBR of the supercardioid beamformer increases at each iteration, and roughly converges after four iterations, while the DF and WNG remain almost the same at each iteration.

5.5.6 Wiener

The MSE between the estimated and desired signals is given by

$$\begin{aligned} J\left(\widetilde{\mathbf{h}}\right) &= E\left(\left|\widetilde{\mathbf{h}}^H \widetilde{\mathbf{y}} - X\right|^2\right) \\ &= \phi_X + \widetilde{\mathbf{h}}^H \boldsymbol{\Phi}_{\widetilde{\mathbf{y}}} \widetilde{\mathbf{h}} - \phi_X \widetilde{\mathbf{h}}^H \widetilde{\mathbf{d}}_0 - \phi_X \widetilde{\mathbf{d}}_0^H \widetilde{\mathbf{h}}, \end{aligned} \tag{5.129}$$

which, thanks to (5.47) and (5.48), can be expressed as

$$J\left(\mathbf{h}_1 \otimes \mathbf{h}_2\right) = \phi_X + \mathbf{h}_1^H \boldsymbol{\Phi}_{\mathbf{y},2} \mathbf{h}_1 - \phi_{X,2} \mathbf{h}_1^H \mathbf{d}_{1,0} - \phi_{X,2}^* \mathbf{d}_{1,0}^H \mathbf{h}_1 \tag{5.130}$$

$$= \phi_X + \mathbf{h}_2^H \boldsymbol{\Phi}_{\mathbf{y},1} \mathbf{h}_2 - \phi_{X,1} \mathbf{h}_2^H \mathbf{d}_{2,0} - \phi_{X,1}^* \mathbf{d}_{2,0}^H \mathbf{h}_2, \tag{5.131}$$

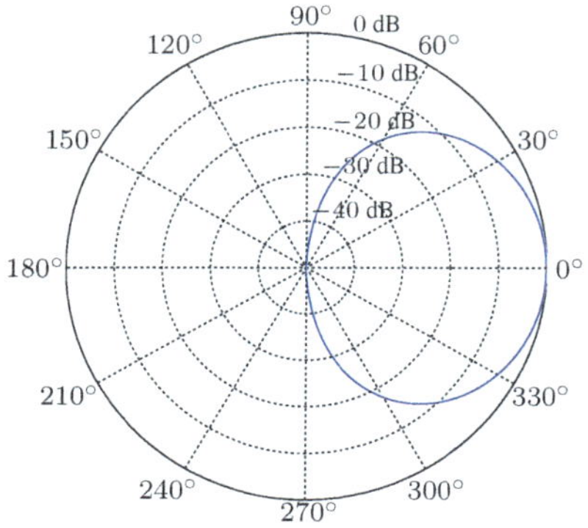

(a)

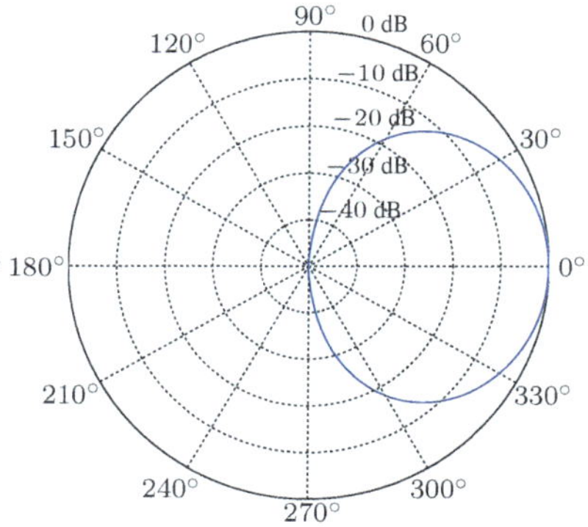

(b)

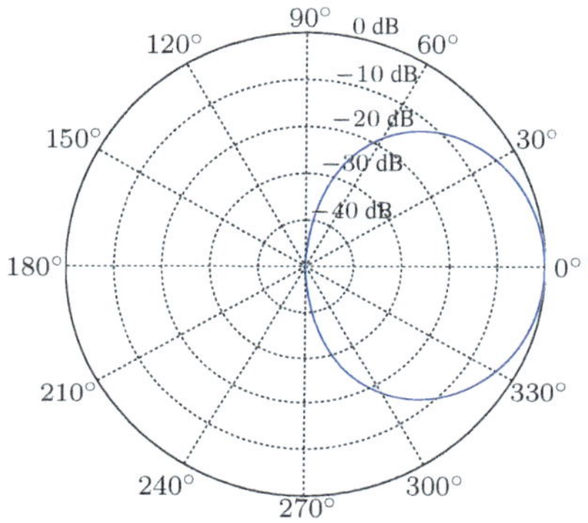

(c)

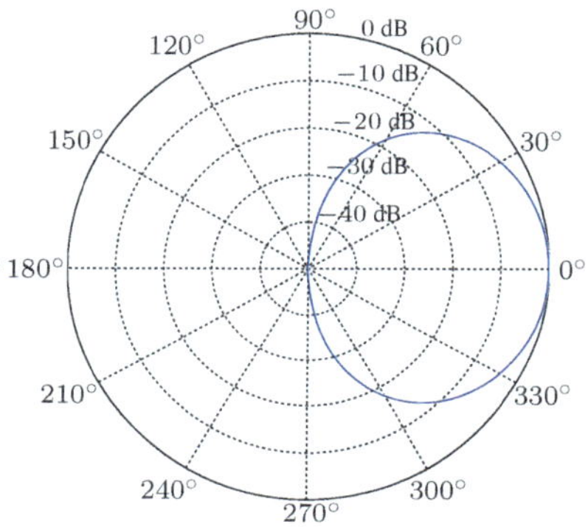

(d)

Fig. 5.13 Beampatterns of the supercardioid beamformer, $\widetilde{\mathbf{h}}_{\mathrm{S}}^{(n)}$, for $f = 2$ kHz, $\delta = 5$ mm, $M_1 = 3$, and $M_2 = 4$, obtained at the iteration n: (a) $n = 1$, (b) $n = 2$, (c) $n = 3$, and (d) $n = 4$.

where

$$\mathbf{\Phi}_{\mathbf{y},2} = \left(\mathbf{I}_{M_1} \otimes \mathbf{h}_2\right)^H \mathbf{\Phi}_{\widetilde{\mathbf{y}}} \left(\mathbf{I}_{M_1} \otimes \mathbf{h}_2\right), \tag{5.132}$$

$$\phi_{X,2} = \phi_X \mathbf{h}_2^H \mathbf{d}_{2,0}, \tag{5.133}$$

and

$$\mathbf{\Phi}_{\mathbf{y},1} = \left(\mathbf{h}_1 \otimes \mathbf{I}_{M_2}\right)^H \mathbf{\Phi}_{\widetilde{\mathbf{y}}} \left(\mathbf{h}_1 \otimes \mathbf{I}_{M_2}\right), \tag{5.134}$$

$$\phi_{X,1} = \phi_X \mathbf{h}_1^H \mathbf{d}_{1,0}. \tag{5.135}$$

It is important to notice that the sizes of the matrices $\mathbf{\Phi}_{\mathbf{y},1}$ and $\mathbf{\Phi}_{\mathbf{y},2}$, which are $M_2 \times M_2$ and $M_1 \times M_1$, respectively, are much smaller than the size of $\mathbf{\Phi}_{\widetilde{\mathbf{y}}}$, which is $M_1 M_2 \times M_1 M_2$. As a result, in practice, much less observations are needed to accurately estimate $\mathbf{\Phi}_{\mathbf{y},1}$ and $\mathbf{\Phi}_{\mathbf{y},2}$ than $\mathbf{\Phi}_{\widetilde{\mathbf{y}}}$, which is the matrix that is inverted in the conventional Wiener beamformer.

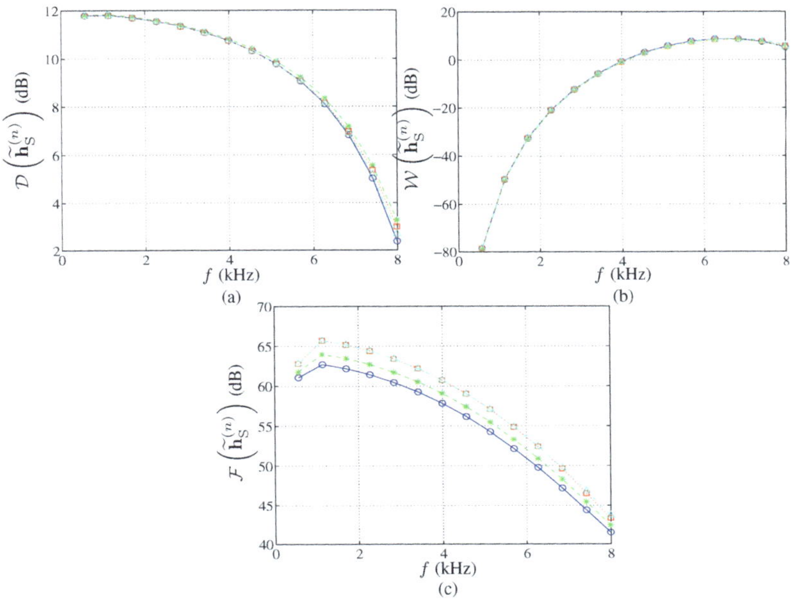

Fig. 5.14 Performance of the supercardioid beamformer, $\widetilde{\mathbf{h}}_{\mathrm{S}}^{(n)}$, as a function of frequency for $\delta = 5$ mm, $M_1 = 3$, $M_2 = 4$, and several values of n: $n = 1$ (solid line with circles), $n = 2$ (dashed line with asterisks), $n = 4$ (dotted line with squares), and $n = 10$ (dash-dot line with triangles). (a) DF, (b) WNG, and (c) FBR.

When $\mathbf{h}_2$ is fixed, we write (5.130) as

$$J(\mathbf{h}_1|\mathbf{h}_2) = \phi_X + \mathbf{h}_1^H \mathbf{\Phi}_{\mathbf{y},2}\mathbf{h}_1 - \phi_{X,2}\mathbf{h}_1^H \mathbf{d}_{1,0} - \phi_{X,2}^* \mathbf{d}_{1,0}^H \mathbf{h}_1, \tag{5.136}$$

and when $\mathbf{h}_1$ is fixed, we write (5.131) as

$$J(\mathbf{h}_2|\mathbf{h}_1) = \phi_X + \mathbf{h}_2^H \mathbf{\Phi}_{\mathbf{y},1}\mathbf{h}_2 - \phi_{X,1}\mathbf{h}_2^H \mathbf{d}_{2,0} - \phi_{X,1}^* \mathbf{d}_{2,0}^H \mathbf{h}_2. \tag{5.137}$$

Now, we have everything to derive an iterative algorithm similar to the one proposed in [11]. At iteration 0, we may take

$$\mathbf{h}_2^{(0)} = \phi_X \mathbf{\Phi}_{\mathbf{y}_2}^{-1} \mathbf{d}_{2,0}, \tag{5.138}$$

where $\mathbf{\Phi}_{\mathbf{y}_2}$ is the covariance matrix of $\mathbf{y}_2$, which is the observation vector corresponding to the second virtual array. In fact, $\mathbf{h}_2^{(0)}$ is just the traditional Wiener beamformer applied to the second virtual ULA. Substituting $\mathbf{h}_2^{(0)}$ into (5.132)–(5.133), we get

$$\mathbf{\Phi}_{\mathbf{y},2}^{(0)} = \left(\mathbf{I}_{M_1} \otimes \mathbf{h}_2^{(0)}\right)^H \mathbf{\Phi}_{\widetilde{\mathbf{y}}} \left(\mathbf{I}_{M_1} \otimes \mathbf{h}_2^{(0)}\right), \tag{5.139}$$

$$\phi_{X,2}^{(0)} = \phi_X \left(\mathbf{h}_2^{(0)}\right)^H \mathbf{d}_{2,0}. \tag{5.140}$$

Then, substituting these quantities into the MSE in (5.136), we obtain at iteration 1:

$$J\left(\mathbf{h}_1^{(1)}|\mathbf{h}_2^{(0)}\right) = \phi_X + \left(\mathbf{h}_1^{(1)}\right)^H \mathbf{\Phi}_{\mathbf{y},2}^{(0)}\mathbf{h}_1^{(1)} - \phi_{X,2}^{(0)} \left(\mathbf{h}_1^{(1)}\right)^H \mathbf{d}_{1,0} - \left(\phi_{X,2}^{(0)}\right)^* \mathbf{d}_{1,0}^H\mathbf{h}_1^{(1)}. \tag{5.141}$$

The minimization of $J\left(\mathbf{h}_1^{(1)}|\mathbf{h}_2^{(0)}\right)$ with respect to $\mathbf{h}_1^{(1)}$ gives

$$\mathbf{h}_1^{(1)} = \phi_{X,2}^{(0)} \left(\mathbf{\Phi}_{\mathbf{y},2}^{(0)}\right)^{-1} \mathbf{d}_{1,0}. \tag{5.142}$$

Using $\mathbf{h}_1^{(1)}$ into (5.134)–(5.135), we obtain

$$\mathbf{\Phi}_{\mathbf{y},1}^{(1)} = \left(\mathbf{h}_1^{(1)} \otimes \mathbf{I}_{M_2}\right)^H \mathbf{\Phi}_{\widetilde{\mathbf{y}}} \left(\mathbf{h}_1^{(1)} \otimes \mathbf{I}_{M_2}\right), \tag{5.143}$$

$$\phi_{X,1}^{(1)} = \phi_X \left(\mathbf{h}_1^{(1)}\right)^H \mathbf{d}_{1,0}. \tag{5.144}$$

With $\mathbf{\Phi}_{\mathbf{y},1}^{(1)}$ and $\phi_{X,1}^{(1)}$, we can compute the MSE in (5.137) as

$$J\left(\mathbf{h}_2^{(1)}|\mathbf{h}_1^{(1)}\right) = \phi_X + \left(\mathbf{h}_2^{(1)}\right)^H \mathbf{\Phi}_{\mathbf{y},1}^{(1)}\mathbf{h}_2^{(1)} - \phi_{X,1}^{(1)} \left(\mathbf{h}_2^{(1)}\right)^H \mathbf{d}_{2,0} - \left(\phi_{X,1}^{(1)}\right)^* \mathbf{d}_{2,0}^H\mathbf{h}_2^{(1)}, \tag{5.145}$$

whose minimization with respect to $\mathbf{h}_2^{(1)}$ gives

$$\mathbf{h}_2^{(1)} = \phi_{X,1}^{(1)} \left(\mathbf{\Phi}_{\mathbf{y},1}^{(1)}\right)^{-1} \mathbf{d}_{2,0}. \tag{5.146}$$

Continuing the iterations up to the iteration n, we easily get the estimate of the first beamformer:

$$\mathbf{h}_1^{(n)} = \phi_{X,2}^{(n-1)} \left(\mathbf{\Phi}_{\mathbf{y},2}^{(n-1)}\right)^{-1} \mathbf{d}_{1,0}, \tag{5.147}$$

where

$$\phi_{X,2}^{(n-1)} = \phi_X \left(\mathbf{h}_2^{(n-1)}\right)^H \mathbf{d}_{2,0}, \tag{5.148}$$

$$\mathbf{\Phi}_{\mathbf{y},2}^{(n-1)} = \left(\mathbf{I}_{M_1} \otimes \mathbf{h}_2^{(n-1)}\right)^H \mathbf{\Phi}_{\widetilde{\mathbf{y}}} \left(\mathbf{I}_{M_1} \otimes \mathbf{h}_2^{(n-1)}\right), \tag{5.149}$$

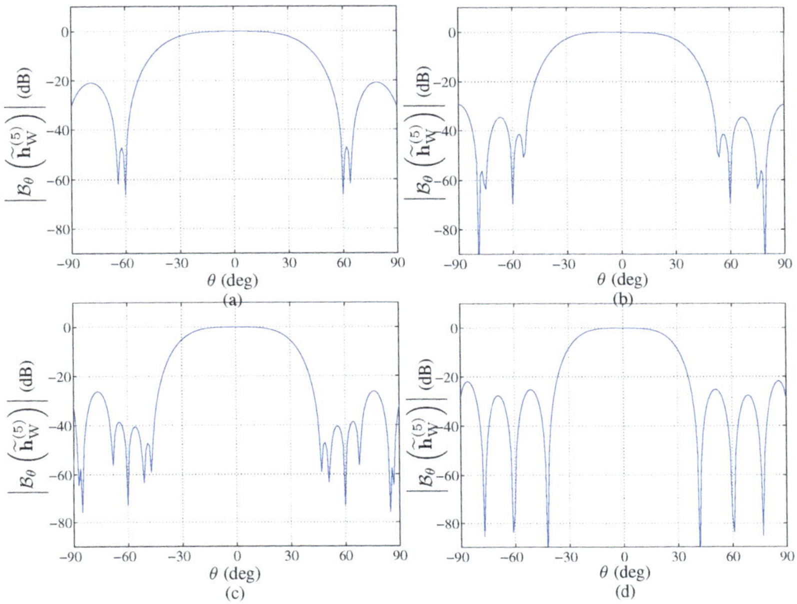

Fig. 5.15 Beampatterns of the Wiener beamformer at the iteration $n = 5$, $\widetilde{\mathbf{h}}_{\mathrm{W}}^{(5)}$, for iSNR = 0 dB, $f = 3$ kHz, $\delta = 1$ cm, $\theta_{\mathrm{d}} = 0°$, $\theta_1 = 60°$, $M_1 = 4$, and different numbers of sensors M_2: (a) $M_2 = 5$, (b) $M_2 = 7$, (c) $M_2 = 9$, and (d) $M_2 = 11$.

and the estimate of the second beamformer:

$$\mathbf{h}_2^{(n)} = \phi_{X,1}^{(n)} \left(\mathbf{\Phi}_{\mathbf{y},1}^{(n)} \right)^{-1} \mathbf{d}_{2,0}, \tag{5.150}$$

where

$$\phi_{X,1}^{(n)} = \phi_X \left(\mathbf{h}_1^{(n)} \right)^H \mathbf{d}_{1,0}, \tag{5.151}$$

$$\mathbf{\Phi}_{\mathbf{y},1}^{(n)} = \left(\mathbf{h}_1^{(n)} \otimes \mathbf{I}_{M_2} \right)^H \mathbf{\Phi}_{\widetilde{\mathbf{y}}} \left(\mathbf{h}_1^{(n)} \otimes \mathbf{I}_{M_2} \right). \tag{5.152}$$

Finally, we deduce that the Wiener beamformer is at iteration n:

$$\widetilde{\mathbf{h}}_{\mathrm{W}}^{(n)} = \mathbf{h}_1^{(n)} \otimes \mathbf{h}_2^{(n)}, \tag{5.153}$$

where $\mathbf{h}_1^{(n)}$ and $\mathbf{h}_2^{(n)}$ are defined in (5.147) and (5.150), respectively.

Example 5.1. Suppose that a desired signal impinges on the ULA from the direction $\theta_{\mathrm{d}} = 0$, and that a statistically independent interference impinges on the ULA from the direction θ_1. Assume that the desired signal is a harmonic pulse of T samples:

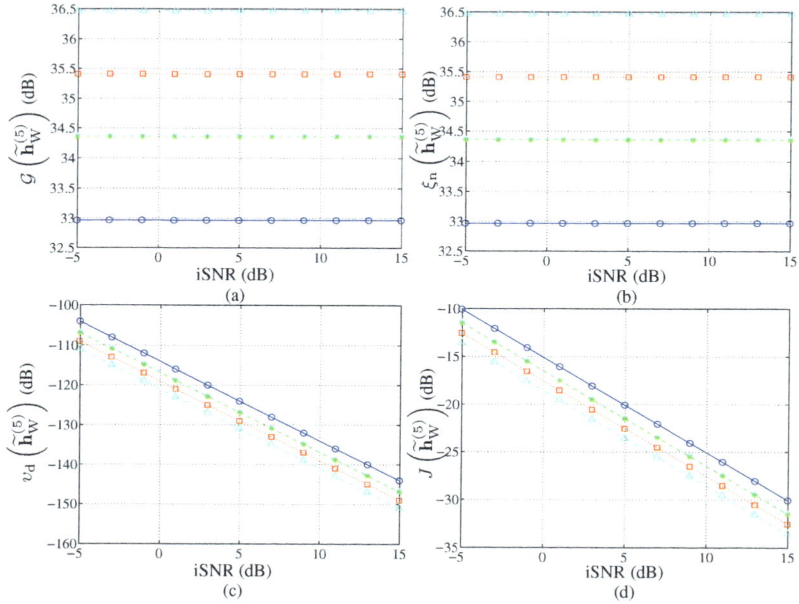

Fig. 5.16 Performance of the Wiener beamformer at the iteration $n = 5$, $\widetilde{\mathbf{h}}_{\mathrm{W}}^{(5)}$, as a function of the input SNR for $f = 3$ kHz, $\delta = 1$ cm, $\theta_{\mathrm{d}} = 0°$, $\theta_1 = 60°$, $M_1 = 4$, and different numbers of sensors M_2: $M_2 = 5$ (solid line with circles), $M_2 = 7$ (dashed line with asterisks), $M_2 = 9$ (dotted line with squares), and $M_2 = 11$ (dash-dot line with triangles). (a) Gain in SNR, (b) noise reduction factor, (c) desired signal distortion index, and (d) MSE.

$$x(t) = \begin{cases} A \sin(\omega_0 t + \phi), & 0 \leq t \leq T - 1 \\ 0, & t < 0,\ t \geq T \end{cases},$$

with fixed amplitude A and angular frequency ω_0, and random phase ϕ, uniformly distributed on the interval from 0 to 2π. Assume that the interference $u(t)$ is white Gaussian noise, i.e., $u(t) \sim \mathcal{N}(0, \sigma_u^2)$, uncorrelated with $x(t)$. In addition, the sensors contain thermal white Gaussian noise, $\widetilde{w}_m(t) \sim \mathcal{N}(0, \sigma_w^2)$, that are mutually uncorrelated. The noisy received signals are given by $\widetilde{y}_m(t) = \widetilde{x}_m(t) + \widetilde{v}_m(t)$, $m = 1, 2, \ldots, M_1 M_2$, where $\widetilde{v}_m(t) = \widetilde{u}_m(t) + \widetilde{w}_m(t)$, $m = 1, 2, \ldots, M_1 M_2$ are the interference-plus-noise signals. The variance of $X(\omega)$ is given by

$$\phi_X = \frac{A^2}{4} D_T^2 [\pi (\omega + \omega_0)] + \frac{A^2}{4} D_T^2 [\pi (\omega - \omega_0)],$$

where

$$D_T(x) = \frac{\sin(Tx)}{\sin(x)}.$$

The covariance matrices of $\widetilde{\mathbf{x}}(\omega)$ and $\widetilde{\mathbf{v}}(\omega)$ are given by

$$\mathbf{\Phi}_{\widetilde{\mathbf{x}}} = \phi_X \widetilde{\mathbf{d}}_0 \widetilde{\mathbf{d}}_0^H,$$
$$\mathbf{\Phi}_{\widetilde{\mathbf{v}}} = T\sigma_u^2 \widetilde{\mathbf{d}}_{\theta_1} \widetilde{\mathbf{d}}_{\theta_1}^H + T\sigma_w^2 \mathbf{I}_M.$$

To demonstrate the performance of the Wiener beamformer, we choose $A = 0.5$, $\omega_0 = 2\pi f_0$, $f_0 = 3$ kH, $T = 500$, $\theta_1 = 60°$, and $\sigma_w^2 = 0.01\sigma_u^2$. Figure 5.15 displays the directivity patterns of the Wiener beamformer at the iteration $n = 5$, for iSNR $= 0$ dB, $f = 3$ kHz, $\delta = 1$ cm, $M_1 = 4$, and different numbers of sensors M_2. As the number of sensors increases, the width of the main beam decreases, and the null in the direction of the interference becomes deeper. Figure 5.16 shows plots of the gain in SNR, $\mathcal{G}\left(\widetilde{\mathbf{h}}_{\mathrm{W}}^{(5)}\right)$, the noise reduction factor, $\xi_{\mathrm{n}}\left(\widetilde{\mathbf{h}}_{\mathrm{W}}^{(5)}\right)$, the desired signal distortion index, $\upsilon_{\mathrm{d}}\left(\widetilde{\mathbf{h}}_{\mathrm{W}}^{(5)}\right)$, and the MSE, $J\left(\widetilde{\mathbf{h}}_{\mathrm{W}}^{(5)}\right)$, as a function of the input SNR for $f = 3$ kHz, $\delta = 1$ cm, $M_1 = 4$, and different numbers of sensors M_2. We observe that as the number of sensors increases, the MSE and the desired signal distortion index obtained by the Wiener beamformer decrease while the gain in SNR and the noise reduction factor increase.

References

1. D. H. Johnson and D. E. Dudgeon, *Array Signal Processing: Concepts and Techniques.* Signal Processing Series. Englewood Cliffs, NJ: Prentice-Hall, 1993.
2. P. P. Vaidyanathan and P. Pal, "Sparse sensing with co-prime samplers and arrays," *IEEE Trans. Signal Process.*, vol. 59, pp. 573–586, Feb. 2011.
3. G. W. Elko and J. Meyer, "Microphone arrays," in *Springer Handbook of Speech Processing*, J. Benesty, M. M. Sondhi, and Y. Huang, Eds., Berlin, Germany: Springer-Verlag, 2008, Chapter 50, pp. 1021–1041.
4. G. W. Elko, "Superdirectional microphone arrays," in *Acoustic Signal Processing for Telecommunication*, S. L. Gay and J. Benesty, Eds. Boston, MA: Kluwer Academic Publishers, 2000, Chapter 10, pp. 181–237.
5. J. Benesty and J. Chen, *Study and Design of Differential Microphone Arrays.* Berlin, Germany: Springer-Verlag, 2012.
6. H. Cox, R. M. Zeskind, and T. Kooij, "Practical supergain," *IEEE Trans. Acoust., Speech, Signal Process.*, vol. ASSP-34, pp. 393–398, June 1986.
7. H. Cox, R. M. Zeskind, and M. M. Owen, "Robust adaptive beamforming," *IEEE Trans. Acoust., Speech, Signal Process.*, vol. ASSP-35, pp. 1365–1376, Oct. 1987.
8. J. Benesty, J. Chen, and Y. Huang, *Microphone Array Signal Processing.* Berlin, Germany: Springer-Verlag, 2008.
9. J. Benesty, I. Cohen, and J. Chen, *Fundamentals of Signal Enhancement and Array Signal Processing.* Singapore: Wiley–IEEE Press, 2018.
10. R. N. Marshall and W. R. Harry, "A new microphone providing uniform directivity over an extended frequency range," *J. Acoust. Soc. Am.*, vol. 12, pp. 481–497, 1941.
11. J. Benesty, C. Paleologu, and S. Ciochina, "On the identification of bilinear forms with the Wiener filter," *IEEE Signal Process. Lett.*, vol. 24, pp. 653–657, May 2017.

Chapter 6
Approach with Rectangular Arrays

All what we have done so far with linear arrays is, obviously, not limited to this type of arrays. In this chapter, we show how to extend some of these results to two-dimensional arrays such as the rectangular ones. The focus is on fixed beamforming. Of course, very interesting adaptive beamformers can be derived as well by following the same steps as in previous chapters.

6.1 Signal Model and Problem Formulation

The two-dimensional (2-D) array evaluated in this chapter is a rectangular array (RA) depicted in Fig. 6.1. Considering the Cartesian coordinate system with microphone $(1, 1)$ as its origin, the studied RA is composed of M_x omnidirectional sensors along the x (negative) axis with a uniform interelement spacing equal to δ_x and M_y omnidirectional sensors along the y (negative) axis with a uniform interelement spacing equal to δ_y. Thus, the total number of microphones is equal to $M_x M_y$, whose positions are denoted (m_x, m_y) with $m_x = 1, 2, \ldots, M_x$, $m_y = 1, 2, \ldots, M_y$. Notice that in the direction of the x axis, we have M_y parallel ULAs composed of M_x microphones each with a spacing δ_x, while in the direction of the y axis, we have M_x parallel ULAs composed of M_y microphones each with a spacing δ_y.

We assume that a farfield desired source signal (plane wave) propagates from the azimuth angle, θ, in an anechoic acoustic environment at the speed of sound, i.e., $c = 340$ m/s, and impinges on the above described 2-D array. Then, it is not hard to see that the corresponding steering matrix (of size $M_x \times M_y$) is [1]

J. Benesty et al., *Array Processing*, Springer Topics in Signal Processing 18, https://doi.org/10.1007/978-3-030-15600-8_6

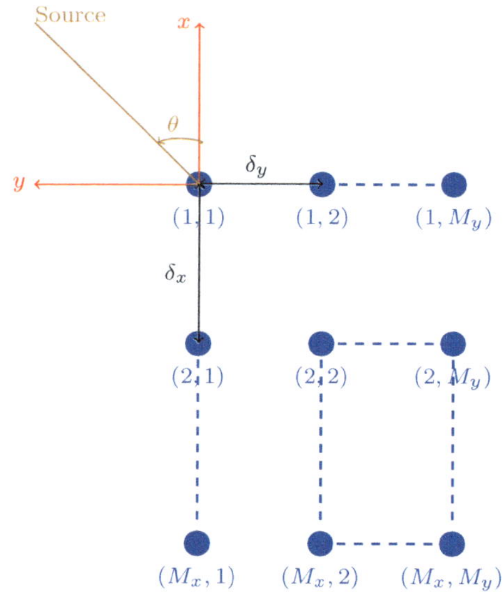

Fig. 6.1 Illustration of the studied 2-D microphone array.

$$\begin{aligned}\mathbf{D}_\theta &= \begin{bmatrix} D_{y,\theta,1}\mathbf{d}_{x,\theta} & D_{y,\theta,2}\mathbf{d}_{x,\theta} & \cdots & D_{y,\theta,M_y}\mathbf{d}_{x,\theta}\end{bmatrix} \\ &= \begin{bmatrix} D_{x,\theta,1}\mathbf{d}_{y,\theta}^T \\ D_{x,\theta,2}\mathbf{d}_{y,\theta}^T \\ \vdots \\ D_{x,\theta,M_x}\mathbf{d}_{y,\theta}^T \end{bmatrix} \\ &= \mathbf{d}_{x,\theta}\mathbf{d}_{y,\theta}^T,\end{aligned} \tag{6.1}$$

where

$$\begin{aligned}\mathbf{d}_{x,\theta} &= \begin{bmatrix} D_{x,\theta,1} & D_{x,\theta,2} & \cdots & D_{x,\theta,M_x}\end{bmatrix}^T \\ &= \begin{bmatrix} 1 & e^{-\jmath\frac{\omega\delta_x}{c}\cos\theta} & \ldots & e^{-\jmath\frac{(M_x-1)\omega\delta_x}{c}\cos\theta}\end{bmatrix}^T\end{aligned} \tag{6.2}$$

is the steering vector associated with the x axis,

$$\begin{aligned}\mathbf{d}_{y,\theta} &= \begin{bmatrix} D_{y,\theta,1} & D_{y,\theta,2} & \cdots & D_{y,\theta,M_y}\end{bmatrix}^T \\ &= \begin{bmatrix} 1 & e^{-\jmath\frac{\omega\delta_y}{c}\sin\theta} & \ldots & e^{-\jmath\frac{(M_y-1)\omega\delta_y}{c}\sin\theta}\end{bmatrix}^T\end{aligned} \tag{6.3}$$

is the steering vector associated with the y axis.

Thanks to the above steering matrix, the observation signal matrix of size $M_x \times M_y$ of the RA can be expressed in the frequency domain as

$$\begin{aligned}\mathbf{Y} &= \mathbf{X} + \mathbf{V} \\ &= \mathbf{D}_\theta X + \mathbf{V},\end{aligned} \tag{6.4}$$

where X is the zero-mean desired source signal and $\mathbf{V}$ is the zero-mean additive noise signal matrix. Without loss of generality, it is assumed in the rest that the desired source signal propagates from the angle $\theta = 0$ (endfire direction). Therefore, (6.4) becomes

$$\mathbf{Y} = \mathbf{D}_0 X + \mathbf{V}, \tag{6.5}$$

where $\mathbf{D}_0$ is the steering matrix at $\theta = 0$.

Using the convenient vectorization operation, which consists of converting a matrix into a vector, (6.5) can be expressed, equivalently, as

$$\begin{aligned}\text{vec}\left(\mathbf{Y}\right) &= \left[\mathbf{y}_{:1}^T \ \mathbf{y}_{:2}^T \ \cdots \ \mathbf{y}_{:M_y}^T\right]^T \\ &= \text{vec}\left(\mathbf{d}_{x,0}\mathbf{d}_{y,0}^T\right) X + \text{vec}\left(\mathbf{V}\right) \\ &= \mathbf{d}_{y,0} \otimes \mathbf{d}_{x,0} X + \text{vec}\left(\mathbf{V}\right),\end{aligned} \tag{6.6}$$

where $\mathbf{y}_{:m_y}$ is the m_yth column of $\mathbf{Y}$ and $\text{vec}\left(\mathbf{V}\right)$ is defined similarly to $\text{vec}\left(\mathbf{Y}\right)$. To further simplify the notation, we write $\widetilde{\mathbf{y}} = \text{vec}\left(\mathbf{Y}\right)$, $\widetilde{\mathbf{d}}_\theta = \text{vec}\left(\mathbf{D}_\theta\right) = \mathbf{d}_{y,\theta} \otimes \mathbf{d}_{x,\theta}$, and $\widetilde{\mathbf{v}} = \text{vec}\left(\mathbf{V}\right)$. With this notation, (6.6) is

$$\widetilde{\mathbf{y}} = \widetilde{\mathbf{d}}_0 X + \widetilde{\mathbf{v}}. \tag{6.7}$$

We deduce that the $M_x M_y \times M_x M_y$ covariance matrix of $\widetilde{\mathbf{y}}$ is

$$\begin{aligned}\mathbf{\Phi}_{\widetilde{\mathbf{y}}} &= E\left(\widetilde{\mathbf{y}}\widetilde{\mathbf{y}}^H\right) \\ &= \widetilde{\mathbf{d}}_0 \widetilde{\mathbf{d}}_0^H \phi_X + \mathbf{\Phi}_{\widetilde{\mathbf{v}}},\end{aligned} \tag{6.8}$$

where $\phi_X = E\left(|X|^2\right)$ is the variance of X and $\mathbf{\Phi}_{\widetilde{\mathbf{v}}} = E\left(\widetilde{\mathbf{v}}\widetilde{\mathbf{v}}^H\right)$ is the covariance matrix of $\widetilde{\mathbf{v}}$.

6.2 2-D Beamforming

The conventional way to perform 2-D beamforming is to apply a complex-valued linear filter, $\mathbf{h}$, of length $M_x M_y$ to the observation signal vector, $\widetilde{\mathbf{y}}$. We get [2]

$$\begin{aligned}Z_\text{C} &= \mathbf{h}^H \widetilde{\mathbf{y}} \\ &= \mathbf{h}^H \widetilde{\mathbf{d}}_0 X + \mathbf{h}^H \widetilde{\mathbf{v}},\end{aligned} \tag{6.9}$$

where Z_{C} is the estimate of the desired signal, X. However, there are two main problems with this approach. The first one is the large number of coefficients (equal to $M_x M_y$) that need to be estimated, so complexity can be an issue. More importantly, the second problem has to do with the inversion of very ill-conditioned large matrices in most derived optimal beamformers, which will necessarily lead to serious estimation problems in the presence of uncertainties.

Due to some potential problems with the conventional approach, we propose in this investigation to use two complex-valued linear filters $\mathbf{h}_x$ and $\mathbf{h}_y$ of respective lengths M_x and M_y as follows[1]:

$$\begin{aligned} Z &= \mathbf{h}_x^H \mathbf{Y} \mathbf{h}_y^* \\ &= \left(\mathbf{h}_x^H \mathbf{d}_{x,0}\right) \left(\mathbf{d}_{y,0}^T \mathbf{h}_y^*\right) X + \mathbf{h}_x^H \mathbf{V} \mathbf{h}_y^* \\ &= \left(\mathbf{h}_x^H \mathbf{d}_{x,0}\right) \left(\mathbf{h}_y^H \mathbf{d}_{y,0}\right) X + \mathbf{h}_x^H \mathbf{V} \mathbf{h}_y^*, \end{aligned} \tag{6.10}$$

where Z is the estimate of X. We observe that Z is bilinear in $\mathbf{h}_x^*$ and $\mathbf{h}_y^*$ since, for every fixed $\mathbf{h}_x^*$, it is a linear function of $\mathbf{h}_y^*$ and for every fixed $\mathbf{h}_y^*$, it is a linear function of $\mathbf{h}_x^*$. This bilinear form takes advantage of the structure of the 2-D array and the corresponding steering vectors. We can also express (6.10) as

$$\begin{aligned} Z &= \mathrm{tr}\left(\mathbf{h}_y^* \mathbf{h}_x^H \mathbf{Y}\right) \\ &= \mathrm{tr}\left[\left(\mathbf{h}_x \mathbf{h}_y^T\right)^H \mathbf{Y}\right] \\ &= \mathrm{vec}^H\left(\mathbf{h}_x \mathbf{h}_y^T\right) \mathrm{vec}\left(\mathbf{Y}\right) \\ &= \left(\mathbf{h}_y \otimes \mathbf{h}_x\right)^H \widetilde{\mathbf{y}}, \end{aligned} \tag{6.11}$$

where $\mathrm{tr}(\cdot)$ denotes the trace of a square matrix and $\mathbf{h}_y \otimes \mathbf{h}_x$ is the global beamformer, which is simply the Kronecker product between the two individual beamformers $\mathbf{h}_y$ and $\mathbf{h}_x$ along the y and x axes, respectively. As a consequence, by taking $\mathbf{h} = \mathbf{h}_y \otimes \mathbf{h}_x$ in (6.9), we observe that Kronecker product beamforming is an interesting particular case of the conventional approach, where $\mathbf{h}$ is assumed to have a particular structure. From (6.11), we find that the variance of Z is

$$\begin{aligned} \phi_Z &= \left(\mathbf{h}_y \otimes \mathbf{h}_x\right)^H \mathbf{\Phi}_{\widetilde{\mathbf{y}}} \left(\mathbf{h}_y \otimes \mathbf{h}_x\right) \\ &= \left|\left(\mathbf{h}_y \otimes \mathbf{h}_x\right)^H \widetilde{\mathbf{d}}_0\right|^2 \phi_X + \left(\mathbf{h}_y \otimes \mathbf{h}_x\right)^H \mathbf{\Phi}_{\widetilde{\mathbf{v}}} \left(\mathbf{h}_y \otimes \mathbf{h}_x\right) \\ &= \left|\mathbf{h}_x^H \mathbf{d}_{x,0}\right|^2 \left|\mathbf{h}_y^H \mathbf{d}_{y,0}\right|^2 \phi_X + \left(\mathbf{h}_y \otimes \mathbf{h}_x\right)^H \mathbf{\Phi}_{\widetilde{\mathbf{v}}} \left(\mathbf{h}_y \otimes \mathbf{h}_x\right). \end{aligned} \tag{6.12}$$

---

[1] Now, the number of coefficients to be estimated is equal to $M_x + M_y$ instead of $M_x M_y$ for the conventional method. If $M_x = M_y = M$, we only need to handle a linear number ($2M$) of coefficients instead of a square number (M^2) of coefficients. So when M is large, the length of the filter in the conventional approach becomes quickly prohibitive.

In the rest, the distortionless constraint is desired, i.e.,

$$\left(\mathbf{h}_x^H \mathbf{d}_{x,0}\right)\left(\mathbf{h}_y^H \mathbf{d}_{y,0}\right) = 1. \tag{6.13}$$

This also means that the value of the Kronecker product beamformer pattern should be equal to 1 at $\theta = 0$ and smaller than 1 at $\theta \neq 0$. In particular, when $\mathbf{h}_x^H \mathbf{d}_{x,0} = \mathbf{h}_y^H \mathbf{d}_{y,0} = 1$, then (6.13) is also verified; so we will always consider this case.

6.3 Performance Measures

The first important measure discussed in this section is the beampattern, which describes the sensitivity of the Kronecker product beamformer to a plane wave impinging on the 2-D array from the direction θ. Mathematically, it is defined as

$$\begin{aligned}
\mathcal{B}_\theta\left(\mathbf{h}_y \otimes \mathbf{h}_x\right) &= \left(\mathbf{h}_y \otimes \mathbf{h}_x\right)^H \widetilde{\mathbf{d}}_\theta \\
&= \left(\mathbf{h}_y \otimes \mathbf{h}_x\right)^H \left(\mathbf{d}_{y,\theta} \otimes \mathbf{d}_{x,\theta}\right) \\
&= \left(\mathbf{h}_x^H \mathbf{d}_{x,\theta}\right)\left(\mathbf{h}_y^H \mathbf{d}_{y,\theta}\right) \\
&= \mathcal{B}_{x,\theta}\left(\mathbf{h}_x\right) \times \mathcal{B}_{y,\theta}\left(\mathbf{h}_y\right).
\end{aligned} \tag{6.14}$$

It is interesting to observe that the beampattern of the global beamformer is equal to the product of the beampatterns of the individual beamformers. As a result, the nulls of $\mathcal{B}_\theta\left(\mathbf{h}_y \otimes \mathbf{h}_x\right)$ correspond exactly to the nulls of $\mathcal{B}_{x,\theta}\left(\mathbf{h}_x\right)$ and $\mathcal{B}_{y,\theta}\left(\mathbf{h}_y\right)$. In particular, if $\mathcal{B}_{x,\theta}\left(\mathbf{h}_x\right)$ has a null at θ_0 of multiplicity 1 and $\mathcal{B}_{y,\theta}\left(\mathbf{h}_y\right)$ has also a null at θ_0 of multiplicity 1, then $\mathcal{B}_\theta\left(\mathbf{h}_y \otimes \mathbf{h}_x\right)$ has a null in the same direction but of multiplicity 2.

Considering the origin of the Cartesian coordinates as the reference, we define the input SNR with respect to this reference as

$$\text{iSNR} = \frac{\phi_X}{\phi_{V_{11}}}, \tag{6.15}$$

where $\phi_{V_{11}}$ is the variance at the noise reference.

The output SNR is defined as

$$\begin{aligned}
\text{oSNR}\left(\mathbf{h}_y \otimes \mathbf{h}_x\right) &= \phi_X \frac{\left|\mathbf{h}_x^H \mathbf{d}_{x,0}\right|^2 \left|\mathbf{h}_y^H \mathbf{d}_{y,0}\right|^2}{\left(\mathbf{h}_y \otimes \mathbf{h}_x\right)^H \mathbf{\Phi}_{\widetilde{\mathbf{v}}} \left(\mathbf{h}_y \otimes \mathbf{h}_x\right)} \\
&= \frac{\phi_X}{\phi_{V_{11}}} \times \frac{\left|\mathbf{h}_x^H \mathbf{d}_{x,0}\right|^2 \left|\mathbf{h}_y^H \mathbf{d}_{y,0}\right|^2}{\left(\mathbf{h}_y \otimes \mathbf{h}_x\right)^H \mathbf{\Gamma}_{\widetilde{\mathbf{v}}} \left(\mathbf{h}_y \otimes \mathbf{h}_x\right)},
\end{aligned} \tag{6.16}$$

where

$$\mathbf{\Gamma}_{\widetilde{\mathbf{v}}} = \frac{\mathbf{\Phi}_{\widetilde{\mathbf{v}}}}{\phi_{V_{11}}} \tag{6.17}$$

is the pseudo-coherence matrix of $\widetilde{\mathbf{v}}$.

The definition of the SNR gain is easily derived from the two previous definitions of the input and output SNRs, i.e.,

$$\begin{aligned} \mathcal{G}\left(\mathbf{h}_y \otimes \mathbf{h}_x\right) &= \frac{\text{oSNR}\left(\mathbf{h}_y \otimes \mathbf{h}_x\right)}{\text{iSNR}} \\ &= \frac{\left|\mathbf{h}_x^H \mathbf{d}_{x,0}\right|^2 \left|\mathbf{h}_y^H \mathbf{d}_{y,0}\right|^2}{\left(\mathbf{h}_y \otimes \mathbf{h}_x\right)^H \mathbf{\Gamma}_{\widetilde{\mathbf{v}}} \left(\mathbf{h}_y \otimes \mathbf{h}_x\right)}. \end{aligned} \tag{6.18}$$

The best known way to evaluate the sensitivity of an array to some of its imperfections and other uncertainties is via the WNG, which is defined by taking $\mathbf{\Gamma}_{\widetilde{\mathbf{v}}} = \mathbf{I}_{M_x M_y}$ in (6.18), where $\mathbf{I}_{M_x M_y}$ is the $M_x M_y \times M_x M_y$ identity matrix, i.e.,

$$\begin{aligned} \mathcal{W}\left(\mathbf{h}_y \otimes \mathbf{h}_x\right) &= \frac{\left|\mathbf{h}_x^H \mathbf{d}_{x,0}\right|^2 \left|\mathbf{h}_y^H \mathbf{d}_{y,0}\right|^2}{\left(\mathbf{h}_y \otimes \mathbf{h}_x\right)^H \left(\mathbf{h}_y \otimes \mathbf{h}_x\right)} \\ &= \frac{\left|\mathbf{h}_x^H \mathbf{d}_{x,0}\right|^2}{\mathbf{h}_x^H \mathbf{h}_x} \times \frac{\left|\mathbf{h}_y^H \mathbf{d}_{y,0}\right|^2}{\mathbf{h}_y^H \mathbf{h}_y} \\ &= \mathcal{W}_x\left(\mathbf{h}_x\right) \times \mathcal{W}_y\left(\mathbf{h}_y\right). \end{aligned} \tag{6.19}$$

Since $\mathcal{W}_x\left(\mathbf{h}_x\right) \leq M_x$ and $\mathcal{W}_y\left(\mathbf{h}_y\right) \leq M_y$, we deduce that $\mathcal{W}\left(\mathbf{h}_y \otimes \mathbf{h}_x\right) \leq M_x M_y$.

Another important measure, which quantifies how the 2-D microphone array performs in the presence of reverberation is the DF. For the spherically isotropic noise field, the definition of the DF is

$$\mathcal{D}\left(\mathbf{h}_y \otimes \mathbf{h}_x\right) = \frac{\left|\mathbf{h}_x^H \mathbf{d}_{x,0}\right|^2 \left|\mathbf{h}_y^H \mathbf{d}_{y,0}\right|^2}{\left(\mathbf{h}_y \otimes \mathbf{h}_x\right)^H \widetilde{\mathbf{\Gamma}} \left(\mathbf{h}_y \otimes \mathbf{h}_x\right)}, \tag{6.20}$$

where

$$\widetilde{\mathbf{\Gamma}} = \begin{bmatrix} \mathbf{\Gamma}_1 & \mathbf{\Gamma}_2 & \cdots & \mathbf{\Gamma}_{M_y-1} & \mathbf{\Gamma}_{M_y} \\ \mathbf{\Gamma}_2 & \mathbf{\Gamma}_1 & \cdots & \mathbf{\Gamma}_{M_y-2} & \mathbf{\Gamma}_{M_y-1} \\ \vdots & \vdots & \ddots & \vdots & \vdots \\ \mathbf{\Gamma}_{M_y-1} & \mathbf{\Gamma}_{M_y-2} & \cdots & \mathbf{\Gamma}_1 & \mathbf{\Gamma}_2 \\ \mathbf{\Gamma}_{M_y} & \mathbf{\Gamma}_{M_y-1} & \cdots & \mathbf{\Gamma}_2 & \mathbf{\Gamma}_1 \end{bmatrix} \tag{6.21}$$

is a symmetric block Toeplitz matrix and the elements of the M_y symmetric Toeplitz matrices $\mathbf{\Gamma}_{m_y}$, $m_y = 1, 2, \ldots, M_y$ (of size $M_x \times M_x$) are given by

$$\left(\mathbf{\Gamma}_{m_y}\right)_{ij} = \text{sinc}\left[\frac{\omega\sqrt{(i-j)^2\delta_x^2 + (m_y-1)^2\,\delta_y^2}}{c}\right], \tag{6.22}$$

with $i, j = 1, 2, \ldots, M_x$ and sinc $x = \sin x/x$. It is clear that $\mathcal{D}\left(\mathbf{h}_y \otimes \mathbf{h}_x\right) \leq \widetilde{\mathbf{d}}_0^H \widetilde{\mathbf{\Gamma}}^{-1} \widetilde{\mathbf{d}}_0$. Using the fact that

$$\begin{aligned}\mathbf{h}_y \otimes \mathbf{h}_x &= \left(\mathbf{h}_y \otimes \mathbf{I}_{M_x}\right)\mathbf{h}_x \\ &= \left(\mathbf{I}_{M_y} \otimes \mathbf{h}_x\right)\mathbf{h}_y,\end{aligned} \tag{6.23}$$

where $\mathbf{I}_{M_x}$ and $\mathbf{I}_{M_y}$ are the identity matrices of sizes $M_x \times M_x$ and $M_y \times M_y$, respectively, we can rewrite the DF as

$$\mathcal{D}\left(\mathbf{h}_y \otimes \mathbf{h}_x\right) = \frac{\left|\mathbf{h}_x^H \mathbf{d}_{x,0}\right|^2 \left|\mathbf{h}_y^H \mathbf{d}_{y,0}\right|^2}{\mathbf{h}_x^H \mathbf{\Gamma}_y \mathbf{h}_x} \tag{6.24}$$

and

$$\mathcal{D}\left(\mathbf{h}_y \otimes \mathbf{h}_x\right) = \frac{\left|\mathbf{h}_x^H \mathbf{d}_{x,0}\right|^2 \left|\mathbf{h}_y^H \mathbf{d}_{y,0}\right|^2}{\mathbf{h}_y^H \mathbf{\Gamma}_x \mathbf{h}_y}, \tag{6.25}$$

where

$$\mathbf{\Gamma}_y = \left(\mathbf{h}_y \otimes \mathbf{I}_{M_x}\right)^H \widetilde{\mathbf{\Gamma}} \left(\mathbf{h}_y \otimes \mathbf{I}_{M_x}\right) \tag{6.26}$$

and

$$\mathbf{\Gamma}_x = \left(\mathbf{I}_{M_y} \otimes \mathbf{h}_x\right)^H \widetilde{\mathbf{\Gamma}} \left(\mathbf{I}_{M_y} \otimes \mathbf{h}_x\right). \tag{6.27}$$

If the filter $\mathbf{h}_y$ is fixed and distortionless, i.e., $\mathbf{h}_y^H \mathbf{d}_{y,0} = 1$, we write (6.24) as

$$\mathcal{D}\left(\mathbf{h}_x | \mathbf{h}_y\right) = \frac{\left|\mathbf{h}_x^H \mathbf{d}_{x,0}\right|^2}{\mathbf{h}_x^H \mathbf{\Gamma}_y \mathbf{h}_x}, \tag{6.28}$$

and if the filter $\mathbf{h}_x$ is fixed and distortionless, i.e., $\mathbf{h}_x^H \mathbf{d}_{x,0} = 1$, we write (6.25) as

$$\mathcal{D}\left(\mathbf{h}_y | \mathbf{h}_x\right) = \frac{\left|\mathbf{h}_y^H \mathbf{d}_{y,0}\right|^2}{\mathbf{h}_y^H \mathbf{\Gamma}_x \mathbf{h}_y}. \tag{6.29}$$

6.4 Fixed Beamformers

There is a myriad of fixed Kronecker product beamformers that we can derive from the proposed approach. Here, we give some relevant examples, which are mostly deduced from the above performance measures.

6.4.1 Delay and Sum

The definition of the WNG with the conventional approach is

$$\mathcal{W}(\mathbf{h}) = \frac{\left|\mathbf{h}^H \widetilde{\mathbf{d}}_0\right|^2}{\mathbf{h}^H \mathbf{h}}, \tag{6.30}$$

whose maximization gives the very well-known DS beamformer:

$$\begin{aligned}\mathbf{h}_{\text{DS}} &= \frac{\widetilde{\mathbf{d}}_0}{\widetilde{\mathbf{d}}_0^H \widetilde{\mathbf{d}}_0} \\ &= \frac{\mathbf{d}_{y,0} \otimes \mathbf{d}_{x,0}}{M_x M_y}\end{aligned} \tag{6.31}$$

and the corresponding WNG is, obviously,

$$\mathcal{W}(\mathbf{h}_{\text{DS}}) = M_x M_y. \tag{6.32}$$

Now, from the maximization of the WNG in (6.19) with respect to $\mathbf{h}_x$ and $\mathbf{h}_y$, we obtain

$$\mathbf{h}_{x,\text{DS}} = \frac{\mathbf{d}_{x,0}}{M_x}, \tag{6.33}$$

$$\mathbf{h}_{y,\text{DS}} = \frac{\mathbf{d}_{y,0}}{M_y}. \tag{6.34}$$

As a result, the global beamformer is

$$\mathbf{h}_{y,\text{DS}} \otimes \mathbf{h}_{x,\text{DS}} = \mathbf{h}_{\text{DS}}, \tag{6.35}$$

showing that the DS beamformers with the conventional and Kronecker product approaches coincide and the number of nulls in the corresponding beampattern is smaller than $M_x + M_y - 2$. For $M_x = M_y = M$, $\mathcal{W}(\mathbf{h}_{y,\text{DS}} \otimes \mathbf{h}_{x,\text{DS}}) = M^2$ with only $2M$ coefficients. This shows how the redundancy in an RA is taken advantage of.

Figure 6.2 displays the directivity patterns of the DS beamformer, $\mathbf{h}_{\text{DS}}$, for $f = 2$ kHz, $\delta_x = \delta_y = 2$ cm and different numbers of sensors $M_x = M_y$.

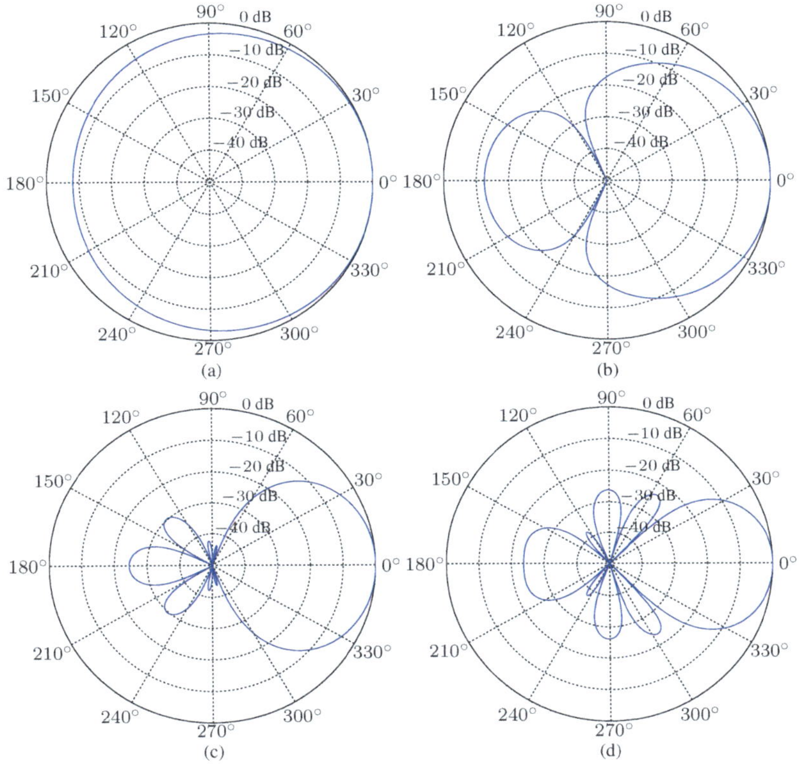

Fig. 6.2 Beampatterns of the DS beamformer, $\mathbf{h}_{\mathrm{DS}}$, for $f = 2$ kHz, $\delta_x = \delta_y = 2$ cm, and different numbers of sensors $M_x = M_y$: (a) $M_x = M_y = 3$, (b) $M_x = M_y = 6$, (c) $M_x = M_y = 9$, and (d) $M_x = M_y = 12$.

Figure 6.3 shows plots of the DFs and WNGs of the DS beamformer as a function of frequency for $\delta_x = \delta_y = 2$ cm and different numbers of sensors $M_x = M_y$. We observe that as the number of sensors increases, the width of the main beam decreases, and both the DF and the WNG of the DS beamformer increase.

6.4.2 Combined Superdirective/Delay and Sum

In this subsection, we show how to combine the superdirective and DS beamformers in a 2-D array in order to take advantage of the best of them. Indeed, for a ULA with the desired source at the endfire, it is well known that the superdirective maximizes the DF but amplifies the white noise while the DS maximizes the WNG but gives poor levels of the DF [3], [4], [5]. Therefore,

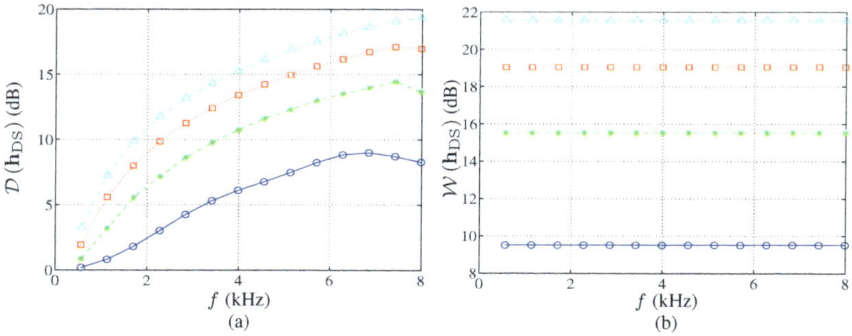

Fig. 6.3 Performance of the DS beamformer, $\mathbf{h}_{\mathrm{DS}}$, as a function of frequency for $\delta_x = \delta_y = 2$ cm and different numbers of sensors $M_x = M_y$: $M_x = M_y = 3$ (solid line with circles), $M_x = M_y = 6$ (dashed line with asterisks), $M_x = M_y = 9$ (dotted line with squares), and $M_x = M_y = 12$ (dash-dot line with triangles). (a) DF and (b) WNG.

in the x axis direction, we propose to use the superdirective beamformer [3], [4]:

$$\mathbf{h}_{x,\mathrm{SD}} = \frac{\mathbf{\Gamma}_1^{-1}\mathbf{d}_{x,0}}{\mathbf{d}_{x,0}^H\mathbf{\Gamma}_1^{-1}\mathbf{d}_{x,0}}, \tag{6.36}$$

where $\mathbf{\Gamma}_1$ is the first block matrix of $\widetilde{\mathbf{\Gamma}}$, while in the y axis direction, we take the DS beamformer given in (6.34), i.e., $\mathbf{h}_{y,\mathrm{DS}}$; so that the global beamformer is $\mathbf{h}_{y,\mathrm{DS}} \otimes \mathbf{h}_{x,\mathrm{SD}}$. We deduce that the WNG is

$$\mathcal{W}\left(\mathbf{h}_{y,\mathrm{DS}} \otimes \mathbf{h}_{x,\mathrm{SD}}\right) = M_y \mathcal{W}_x\left(\mathbf{h}_{x,\mathrm{SD}}\right), \tag{6.37}$$

showing that the WNG of the global filter with an RA is improved by a factor of M_y compared to the WNG of the superdirective beamformer with a ULA. The power beampattern is

$$\begin{aligned} \left|\mathcal{B}_\theta\left(\mathbf{h}_{y,\mathrm{DS}} \otimes \mathbf{h}_{x,\mathrm{SD}}\right)\right|^2 &= \left|\mathcal{B}_{x,\theta}\left(\mathbf{h}_{x,\mathrm{SD}}\right)\right|^2 \left|\mathcal{B}_{y,\theta}\left(\mathbf{h}_{y,\mathrm{DS}}\right)\right|^2 \\ &\leq \left|\mathcal{B}_{x,\theta}\left(\mathbf{h}_{x,\mathrm{SD}}, \theta\right)\right|^2, \end{aligned} \tag{6.38}$$

implying that the global beamformer is more directive than the superdirective beamformer. As a consequence,

$$\mathcal{D}\left(\mathbf{h}_{y,\mathrm{DS}} \otimes \mathbf{h}_{x,\mathrm{SD}}\right) \geq \mathcal{D}\left(\mathbf{h}_{x,\mathrm{SD}}\right). \tag{6.39}$$

Figure 6.4 displays the directivity patterns of the combined superdirective/DS beamformer, $\mathbf{h}_{y,\mathrm{DS}} \otimes \mathbf{h}_{x,\mathrm{SD}}$, for $f = 2$ kHz, $\delta_x = 1$ cm, $\delta_y = 2$ cm, $M_y = 10$, and different numbers of sensors M_x. Figure 6.5 shows plots of the DFs and WNGs of the combined superdirective/DS beamformer as a function of frequency for $\delta_x = 1$ cm, $\delta_y = 2$ cm, $M_y = 10$, and different numbers of

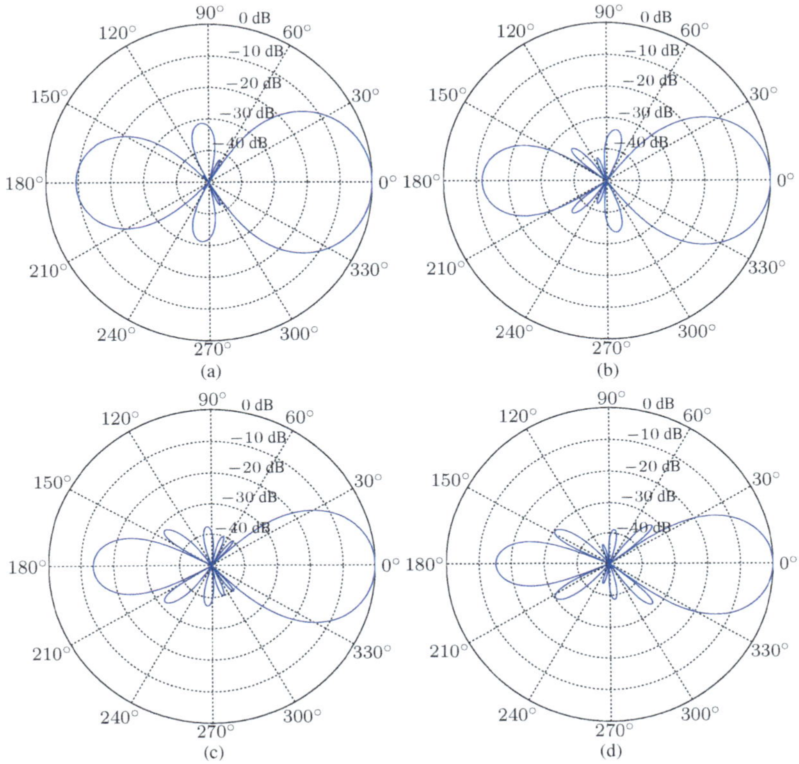

Fig. 6.4 Beampatterns of the combined superdirective/DS beamformer, $\mathbf{h}_{y,\mathrm{DS}} \otimes \mathbf{h}_{x,\mathrm{SD}}$, for $f = 2$ kHz, $\delta_x = 1$ cm, $\delta_y = 2$ cm, $M_y = 10$, and different numbers of sensors M_x: (a) $M_x = 3$, (b) $M_x = 4$, (c) $M_x = 5$, and (d) $M_x = 6$.

sensors M_x. We observe that as the number of sensors increases, the width of the main beam and the level of side lobes decrease, the DF increases, but the WNG decreases. Compared to the DS beamformer, the combined superdirective/DS beamformer yields higher DF, but lower WNG (compare Figs 6.3 and 6.5).

6.4.3 Maximum DF

In the conventional method, the DF is defined as

$$\mathcal{D}(\mathbf{h}) = \frac{\left|\mathbf{h}^H \widetilde{\mathbf{d}}_0\right|^2}{\mathbf{h}^H \widetilde{\mathbf{\Gamma}} \mathbf{h}}. \tag{6.40}$$

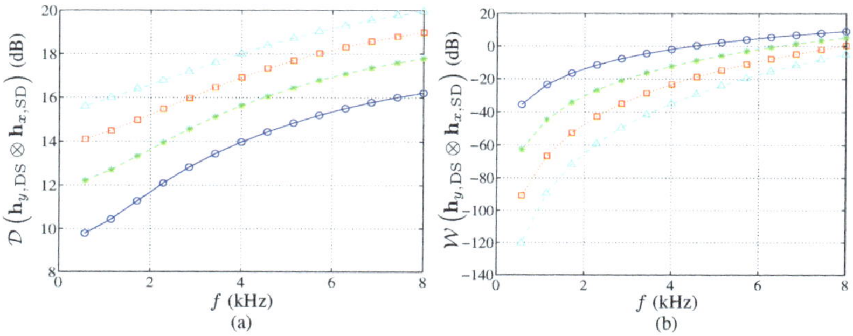

Fig. 6.5 Performance of the combined superdirective/DS beamformer, $\mathbf{h}_{y,\mathrm{DS}} \otimes \mathbf{h}_{x,\mathrm{SD}}$, as a function of frequency for $\delta_x = 1$ cm, $\delta_y = 2$ cm, $M_y = 10$, and different numbers of sensors M_x: $M_x = 3$ (solid line with circles), $M_x = 4$ (dashed line with asterisks), $M_x = 5$ (dotted line with squares), and $M_x = 6$ (dash-dot line with triangles). (a) DF and (b) WNG.

We can easily maximize the previous expression to obtain the maximum DF (mDF) beamformer:

$$\mathbf{h}_{\mathrm{mDF}} = \frac{\widetilde{\mathbf{\Gamma}}^{-1}\widetilde{\mathbf{d}}_0}{\widetilde{\mathbf{d}}_0^H \widetilde{\mathbf{\Gamma}}^{-1}\widetilde{\mathbf{d}}_0}. \tag{6.41}$$

While this approach maximizes the DF, which is equal to $\mathcal{D}(\mathbf{h}_{\mathrm{mDF}}) = \widetilde{\mathbf{d}}_0^H \widetilde{\mathbf{\Gamma}}^{-1}\widetilde{\mathbf{d}}_0$, it may have a disastrous effect on the WNG. Therefore, $\mathbf{h}_{\mathrm{mDF}}$ may be unpractical.

With the Kronecker product technique, it does not seem obvious to maximize the DF [see eq. (6.20)] but we can maximize the DFs in the directions of the two axes x and y. The maximization of the DF of $\mathbf{h}_x$ gives the superdirective beamformer, $\mathbf{h}_{x,\mathrm{SD}}$, shown in (6.36), while the maximization of the DF of $\mathbf{h}_y$ leads to the mDF beamformer:

$$\mathbf{h}_{y,\mathrm{mDF}} = \frac{\mathbf{\Gamma}_{1,y}^{-1}\mathbf{d}_{y,0}}{\mathbf{d}_{y,0}^H \mathbf{\Gamma}_{1,y}^{-1}\mathbf{d}_{y,0}}, \tag{6.42}$$

where the elements of the $M_y \times M_y$ symmetric Toeplitz matrix $\mathbf{\Gamma}_{1,y}$ are given by

$$\left(\mathbf{\Gamma}_{1,y}\right)_{ij} = \mathrm{sinc}\left(\frac{\omega|i-j|\delta_y}{c}\right), \tag{6.43}$$

with $i, j = 1, 2, \ldots, M_y$. Therefore, the global beamformer is $\mathbf{h}_{y,\mathrm{mDF}} \otimes \mathbf{h}_{x,\mathrm{SD}}$, which, obviously, does not maximize (6.20).

We can improve the previous result, as far the DF is concerned, with a simple iterative algorithm thanks to (6.28) and (6.29). At iteration 0, we take

$$\mathbf{h}_x^{(0)} = \mathbf{h}_{x,\mathrm{SD}}, \tag{6.44}$$

where $\mathbf{h}_{x,\mathrm{SD}}$ is given in (6.36). Substituting $\mathbf{h}_x^{(0)}$ into (6.27), we obtain

$$\mathbf{\Gamma}_x^{(0)} = \left(\mathbf{I}_{M_y} \otimes \mathbf{h}_x^{(0)}\right)^H \widetilde{\mathbf{\Gamma}} \left(\mathbf{I}_{M_y} \otimes \mathbf{h}_x^{(0)}\right). \tag{6.45}$$

Now, substituting this expression into the DF in (6.29), we obtain at iteration 1:

$$\mathcal{D}\left(\mathbf{h}_y^{(1)} | \mathbf{h}_x^{(0)}\right) = \frac{\left|\left(\mathbf{h}_y^{(1)}\right)^H \mathbf{d}_{y,0}\right|^2}{\left(\mathbf{h}_y^{(1)}\right)^H \mathbf{\Gamma}_x^{(0)} \mathbf{h}_y^{(1)}}. \tag{6.46}$$

The maximization of $\mathcal{D}\left(\mathbf{h}_y^{(1)} | \mathbf{h}_x^{(0)}\right)$ with respect to $\mathbf{h}_y^{(1)}$ gives

$$\mathbf{h}_y^{(1)} = \frac{\left(\mathbf{\Gamma}_x^{(0)}\right)^{-1} \mathbf{d}_{y,0}}{\mathbf{d}_{y,0}^H \left(\mathbf{\Gamma}_x^{(0)}\right)^{-1} \mathbf{d}_{y,0}}. \tag{6.47}$$

Using $\mathbf{h}_y^{(1)}$ in (6.26), we get

$$\mathbf{\Gamma}_y^{(1)} = \left(\mathbf{h}_y^{(1)} \otimes \mathbf{I}_{M_x}\right)^H \widetilde{\mathbf{\Gamma}} \left(\mathbf{h}_y^{(1)} \otimes \mathbf{I}_{M_x}\right). \tag{6.48}$$

As a result, the DF in (6.28) is

$$\mathcal{D}\left(\mathbf{h}_x^{(1)} | \mathbf{h}_y^{(1)}\right) = \frac{\left|\left(\mathbf{h}_x^{(1)}\right)^H \mathbf{d}_{x,0}\right|^2}{\left(\mathbf{h}_x^{(1)}\right)^H \mathbf{\Gamma}_y^{(1)} \mathbf{h}_x^{(1)}}, \tag{6.49}$$

whose maximization with respect to $\mathbf{h}_x^{(1)}$ gives

$$\mathbf{h}_x^{(1)} = \frac{\left(\mathbf{\Gamma}_y^{(1)}\right)^{-1} \mathbf{d}_{x,0}}{\mathbf{d}_{x,0}^H \left(\mathbf{\Gamma}_y^{(1)}\right)^{-1} \mathbf{d}_{x,0}}. \tag{6.50}$$

Continuing the iterations up to the iteration n, we easily get for the second filter:

$$\mathbf{h}_y^{(n)} = \frac{\left(\mathbf{\Gamma}_x^{(n-1)}\right)^{-1} \mathbf{d}_{y,0}}{\mathbf{d}_{y,0}^H \left(\mathbf{\Gamma}_x^{(n-1)}\right)^{-1} \mathbf{d}_{y,0}}, \tag{6.51}$$

with

$$\mathbf{\Gamma}_x^{(n-1)} = \left(\mathbf{I}_{M_y} \otimes \mathbf{h}_x^{(n-1)}\right)^H \widetilde{\mathbf{\Gamma}} \left(\mathbf{I}_{M_y} \otimes \mathbf{h}_x^{(n-1)}\right), \tag{6.52}$$

and for the first filter:

$$\mathbf{h}_x^{(n)} = \frac{\left(\mathbf{\Gamma}_y^{(n)}\right)^{-1} \mathbf{d}_{x,0}}{\mathbf{d}_{x,0}^H \left(\mathbf{\Gamma}_y^{(n)}\right)^{-1} \mathbf{d}_{x,0}}, \tag{6.53}$$

with

$$\mathbf{\Gamma}_y^{(n)} = \left(\mathbf{h}_y^{(n)} \otimes \mathbf{I}_{M_x}\right)^H \widetilde{\mathbf{\Gamma}} \left(\mathbf{h}_y^{(n)} \otimes \mathbf{I}_{M_x}\right). \tag{6.54}$$

As a result, the global beamformer at iteration n is $\mathbf{h}_y^{(n)} \otimes \mathbf{h}_x^{(n)}$. Since the DFs of the individual beamformers increase at each iteration, so is the DF of the global beamformer and we should expect (6.20) to be maximized for n large enough. While this iterative algorithm may lead to a high value of the DF, white noise amplification may be a serious issue.

Figure 6.6 displays the directivity patterns of the global mDF beamformer at the iteration $n = 5$, $\mathbf{h}_y^{(5)} \otimes \mathbf{h}_x^{(5)}$, for $f = 2$ kHz, $\delta_x = \delta_y = 1$ cm, $M_y = 3$, and different numbers of sensors M_x. Figure 6.7 shows plots of the DFs and WNGs of the global mDF beamformer for $\delta_x = \delta_y = 1$ cm, $M_y = 3$, and different numbers of sensors M_x. We observe that as the number of sensors increases, the DF of the global mDF beamformer increases, but the WNG decreases. Compared to the DS beamformer and the combined superdirective/DS beamformer, the global mDF beamformer yields higher DF, but lower WNG (compare Figs 6.3, 6.5, and 6.7).

One obvious way to better compromise between the DF and the WNG is to change the initialization to

$$\mathbf{h}_x^{(0)} = \mathbf{h}_{x,\mathrm{DS}}, \tag{6.55}$$

where $\mathbf{h}_{x,\mathrm{DS}}$ is the individual DS beamformer defined in (6.33). We can iterate as above using (6.28) and (6.29) and stop when we achieve a desired DF or when we do not desire to go below a certain level of the WNG.

6.4.4 Null Steering

In this subsection, we assume that we have one interference source impinging on the array from the direction $\theta_0 \neq 0$ that we would like to completely cancel, i.e., to steer a null in that direction, and, meanwhile, recover the desired source coming from the endfire direction. With the conventional approach,

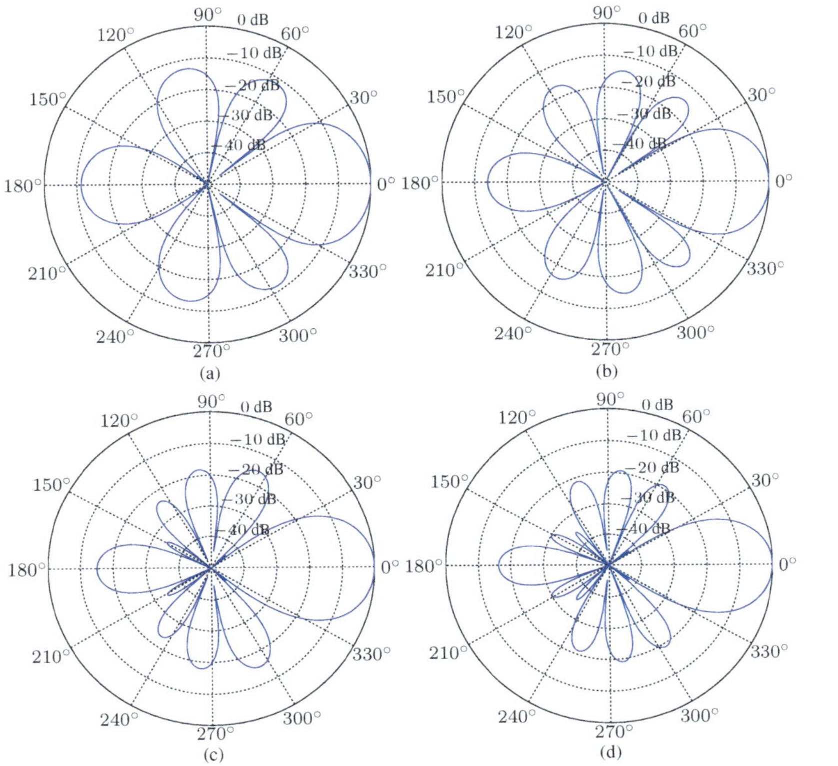

Fig. 6.6 Beampatterns of the global mDF beamformer at the iteration $n = 5$, $\mathbf{h}_y^{(5)} \otimes \mathbf{h}_x^{(5)}$, for $f = 2$ kHz, $\delta_x = \delta_y = 1$ cm, $M_y = 3$, and different numbers of sensors M_x: (a) $M_x = 3$, (b) $M_x = 4$, (c) $M_x = 5$, and (d) $M_x = 6$.

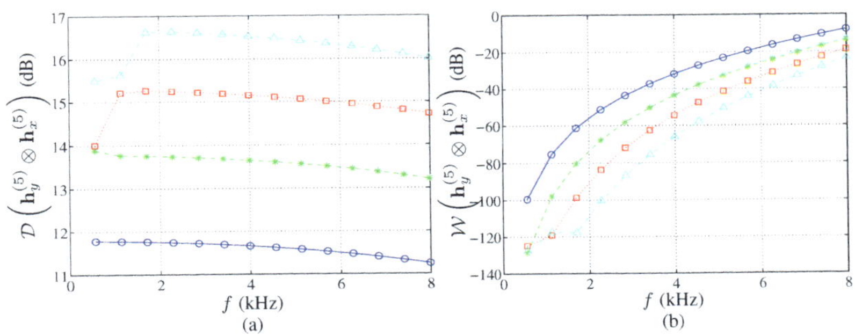

Fig. 6.7 Performance of the global mDF beamformer at the iteration $n = 5$, $\mathbf{h}_y^{(5)} \otimes \mathbf{h}_x^{(5)}$, as a function of frequency for $\delta_x = \delta_y = 1$ cm, $M_y = 3$, and different numbers of sensors M_x: $M_x = 3$ (solid line with circles), $M_x = 4$ (dashed line with asterisks), $M_x = 5$ (dotted line with squares), and $M_x = 6$ (dash-dot line with triangles). (a) DF and (b) WNG.

the constraint equation for a distortionless response and a null at θ_0 is

$$\widetilde{\mathbf{C}}^H \mathbf{h} = \mathbf{i}_{\mathrm{c}}, \tag{6.56}$$

where

$$\widetilde{\mathbf{C}} = \left[\widetilde{\mathbf{d}}_0 \; \widetilde{\mathbf{d}}_{\theta_0} \right] \tag{6.57}$$

is the constraint matrix of size $M_x M_y \times 2$ whose two columns are linearly independent and

$$\mathbf{i}_{\mathrm{c}} = \left[1 \; 0 \right]^T \tag{6.58}$$

is a vector of length 2. There are two interesting ways to find $\mathbf{h}$. The first obvious beamformer is obtained by maximizing the WNG and by taking (6.56) into account, i.e.,

$$\min_{\mathbf{h}} \mathbf{h}^H \mathbf{h} \quad \text{subject to} \quad \widetilde{\mathbf{C}}^H \mathbf{h} = \mathbf{i}_{\mathrm{c}}. \tag{6.59}$$

From this criterion, we find the minimum-norm (MN) beamformer:

$$\mathbf{h}_{\mathrm{MN}} = \widetilde{\mathbf{C}} \left(\widetilde{\mathbf{C}}^H \widetilde{\mathbf{C}} \right)^{-1} \mathbf{i}_{\mathrm{c}}, \tag{6.60}$$

which is also the minimum-norm solution of (6.56). The other beamformer is obtained by maximizing the DF and by taking (6.56) into account, i.e.,

$$\min_{\mathbf{h}} \mathbf{h}^H \widetilde{\mathbf{\Gamma}} \mathbf{h} \quad \text{subject to} \quad \widetilde{\mathbf{C}}^H \mathbf{h} = \mathbf{i}_{\mathrm{c}}. \tag{6.61}$$

We easily find the null-steering (NS) beamformer:

$$\mathbf{h}_{\mathrm{NS}} = \widetilde{\mathbf{\Gamma}}^{-1} \widetilde{\mathbf{C}} \left(\widetilde{\mathbf{C}}^H \widetilde{\mathbf{\Gamma}}^{-1} \widetilde{\mathbf{C}} \right)^{-1} \mathbf{i}_{\mathrm{c}}. \tag{6.62}$$

Obviously, we always have

$$\mathcal{W}\left(\mathbf{h}_{\mathrm{NS}}\right) \leq \mathcal{W}\left(\mathbf{h}_{\mathrm{MN}}\right), \tag{6.63}$$

$$\mathcal{D}\left(\mathbf{h}_{\mathrm{NS}}\right) \geq \mathcal{D}\left(\mathbf{h}_{\mathrm{MN}}\right). \tag{6.64}$$

For the Kronecker product approach, there are many possibilities. Now, for our above formulated problem, the corresponding constraint equations on the two filters $\mathbf{h}_x$ and $\mathbf{h}_y$ are

$$\mathbf{C}_x^H \mathbf{h}_x = \mathbf{i}_{\mathrm{c}}, \tag{6.65}$$

$$\mathbf{C}_y^H \mathbf{h}_y = \mathbf{i}_{\mathrm{c}}, \tag{6.66}$$

where

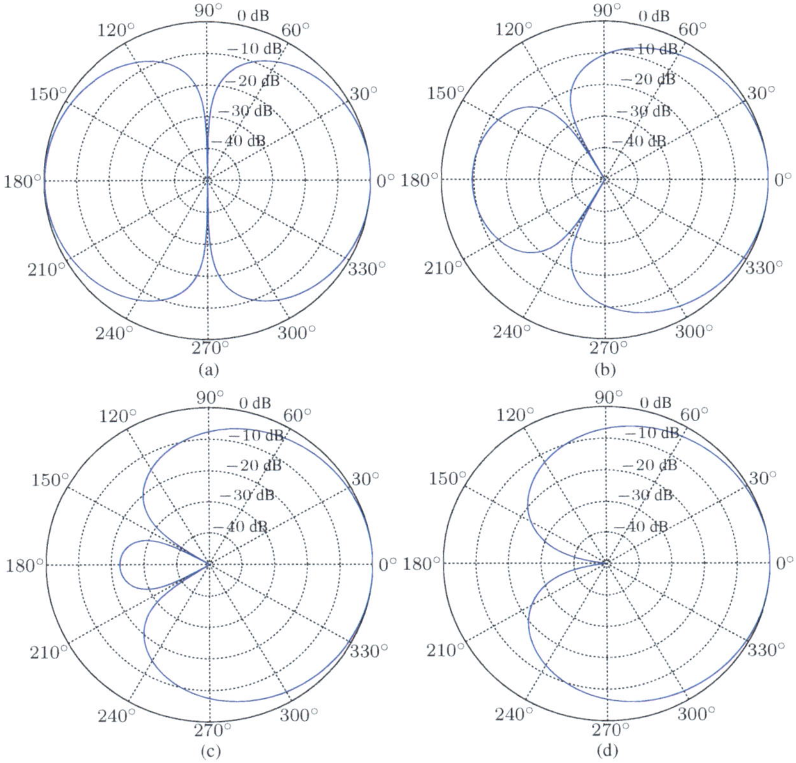

Fig. 6.8 Beampatterns of the global MN beamformer, $\mathbf{h}_{y,\text{DS}} \otimes \mathbf{h}_{x,\text{MN}}$, for $f = 2$ kHz, $\delta_x = \delta_y = 1$ cm, $M_x = 3$, $M_y = 5$, and several values of θ_0: (a) $\theta_0 = 90°$, (b) $\theta_0 = 120°$, (c) $\theta_0 = 150°$, and (d) $\theta_0 = 180°$.

$$\mathbf{C}_x = \left[\mathbf{d}_{x,0} \ \mathbf{d}_{x,\theta_0} \right], \tag{6.67}$$

$$\mathbf{C}_y = \left[\mathbf{d}_{y,0} \ \mathbf{d}_{y,\theta_0} \right] \tag{6.68}$$

are the constraint matrices of size $M_x \times 2$ and $M_y \times 2$, respectively. From (6.65) and (6.66), we easily find the individual MN beamformers:

$$\mathbf{h}_{x,\text{MN}} = \mathbf{C}_x \left(\mathbf{C}_x^H \mathbf{C}_x \right)^{-1} \mathbf{i}_c, \tag{6.69}$$

$$\mathbf{h}_{y,\text{MN}} = \mathbf{C}_y \left(\mathbf{C}_y^H \mathbf{C}_y \right)^{-1} \mathbf{i}_c. \tag{6.70}$$

As a consequence, for the global MN beamformer, we have three interesting possibilities: $\mathbf{h}_{y,\text{DS}} \otimes \mathbf{h}_{x,\text{MN}}$, $\mathbf{h}_{y,\text{MN}} \otimes \mathbf{h}_{x,\text{DS}}$, and $\mathbf{h}_{y,\text{MN}} \otimes \mathbf{h}_{x,\text{MN}}$. The three of them, obviously, put a null in the direction θ_0, but for the last one, the null is of multiplicity 2, and its corresponding WNG (resp. DF) should be smaller (resp. greater) than the two others.

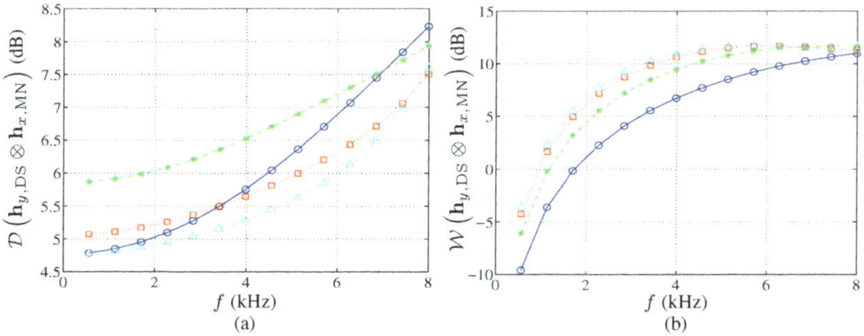

Fig. 6.9 Performance of the global MN beamformer, $\mathbf{h}_{y,\mathrm{DS}} \otimes \mathbf{h}_{x,\mathrm{MN}}$, as a function of frequency for $\delta_x = \delta_y = 1$ cm, $M_x = 3$, $M_y = 5$, and several values of θ_0: $\theta_0 = 90°$ (solid line with circles), $\theta_0 = 120°$ (dashed line with asterisks), $\theta_0 = 150°$ (dotted line with squares), and $\theta_0 = 180°$ (dash-dot line with triangles). (a) DF and (b) WNG.

Figure 6.8 displays the directivity patterns of the global MN beamformer, $\mathbf{h}_{y,\mathrm{DS}} \otimes \mathbf{h}_{x,\mathrm{MN}}$, for $f = 2$ kHz, $\delta_x = \delta_y = 1$ cm, $M_x = 3$, $M_y = 5$, and several values of θ_0. Figure 6.9 shows plots of the DFs and WNGs of the global MN beamformer as a function of frequency for $\delta_x = \delta_y = 1$ cm, $M_x = 3$, $M_y = 5$, and several values of θ_0. We observe a null in the direction θ_0, and the WNG of the global MN beamformer increases as θ_0 increases from 90° to 180°.

Following the steps of the conventional approach, we easily find the individual NS beamformers:

$$\mathbf{h}_{x,\mathrm{NS}} = \mathbf{\Gamma}_1^{-1}\mathbf{C}_x\left(\mathbf{C}_x^H\mathbf{\Gamma}_1^{-1}\mathbf{C}_x\right)^{-1}\mathbf{i}_\mathrm{c}, \tag{6.71}$$

$$\mathbf{h}_{y,\mathrm{NS}} = \mathbf{\Gamma}_{1,y}^{-1}\mathbf{C}_y\left(\mathbf{C}_y^H\mathbf{\Gamma}_{1,y}^{-1}\mathbf{C}_y\right)^{-1}\mathbf{i}_\mathrm{c}, \tag{6.72}$$

and for the global beamformer, we have many more possibilities depending on what we want. Here are some examples: $\mathbf{h}_{y,\mathrm{NS}} \otimes \mathbf{h}_{x,\mathrm{SD}}$, $\mathbf{h}_{y,\mathrm{NS}} \otimes \mathbf{h}_{x,\mathrm{MN}}$, $\mathbf{h}_{y,\mathrm{mDF}} \otimes \mathbf{h}_{x,\mathrm{NS}}$, $\mathbf{h}_{y,\mathrm{MN}} \otimes \mathbf{h}_{x,\mathrm{NS}}$, $\mathbf{h}_{y,\mathrm{DS}} \otimes \mathbf{h}_{x,\mathrm{NS}}$, and $\mathbf{h}_{y,\mathrm{NS}} \otimes \mathbf{h}_{x,\mathrm{NS}}$. The last one will give the best DF.

Figure 6.10 displays the global NS beamformer, $\mathbf{h}_{y,\mathrm{DS}} \otimes \mathbf{h}_{x,\mathrm{NS}}$, for $f = 2$ kHz, $\delta_x = \delta_y = 1$ cm, $M_x = 3$, $M_y = 5$, and several values of θ_0. Figure 6.11 shows plots of the DFs and WNGs of the global NS beamformer, $\mathbf{h}_{y,\mathrm{DS}} \otimes \mathbf{h}_{x,\mathrm{NS}}$, as a function of frequency for $\delta_x = \delta_y = 1$ cm, $M_x = 3$, $M_y = 5$, and several values of θ_0. Figure 6.12 displays the global NS beamformer, $\mathbf{h}_{y,\mathrm{NS}} \otimes \mathbf{h}_{x,\mathrm{NS}}$, for $f = 2$ kHz, $\delta_x = \delta_y = 1$ cm, $M_x = 3$, $M_y = 5$, and several values of θ_0. Figure 6.13 shows plots of the DFs and WNGs of the global NS beamformer, $\mathbf{h}_{y,\mathrm{NS}} \otimes \mathbf{h}_{x,\mathrm{NS}}$, as a function of frequency for $\delta_x = \delta_y = 1$ cm, $M_x = 3$, $M_y = 5$, and several values of θ_0. We observe a null in the direction θ_0. Compared to the global MN beamformers, the global NS beamformers yield higher DF, but lower WNG. The beamformer $\mathbf{h}_{y,\mathrm{NS}} \otimes \mathbf{h}_{x,\mathrm{NS}}$ yields the highest DF, but lowest WNG (compare Figs 6.9, 6.11, and 6.13).

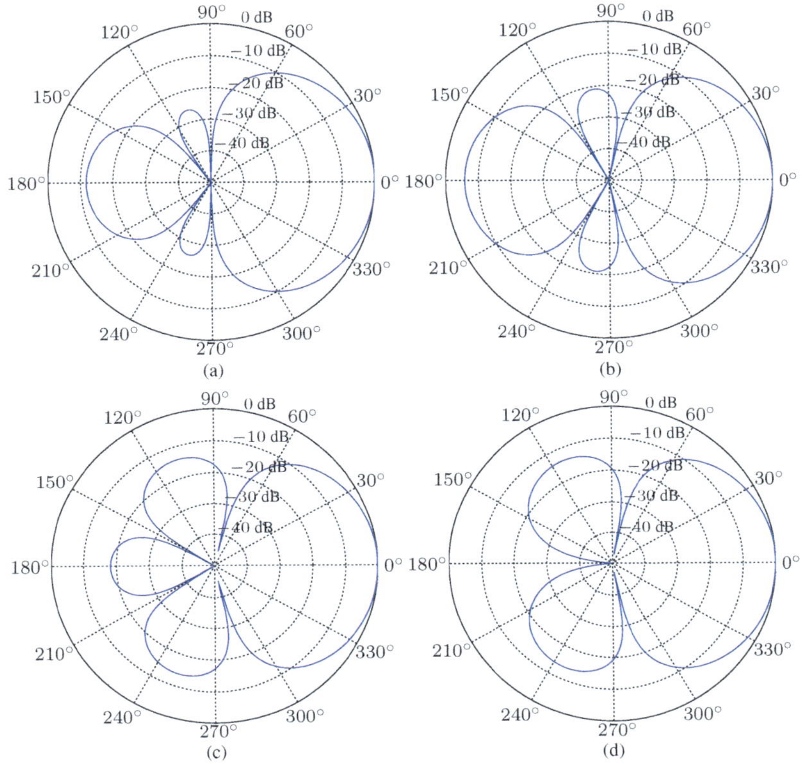

Fig. 6.10 Beampatterns of the global NS beamformer, $\mathbf{h}_{y,\mathrm{DS}} \otimes \mathbf{h}_{x,\mathrm{NS}}$, for $f = 2$ kHz, $\delta_x = \delta_y = 1$ cm, $M_x = 3$, $M_y = 5$, and several values of θ_0: (a) $\theta_0 = 90°$, (b) $\theta_0 = 120°$, (c) $\theta_0 = 150°$, and (d) $\theta_0 = 180°$.

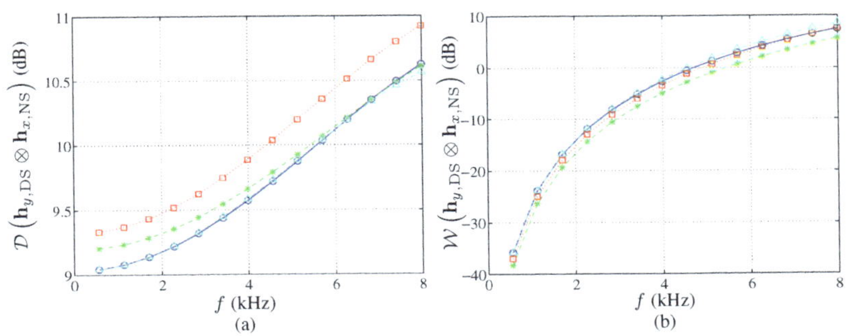

Fig. 6.11 Performance of the global NS beamformer, $\mathbf{h}_{y,\mathrm{DS}} \otimes \mathbf{h}_{x,\mathrm{NS}}$, as a function of frequency for $\delta_x = \delta_y = 1$ cm, $M_x = 3$, $M_y = 5$, and several values of θ_0: $\theta_0 = 90°$ (solid line with circles), $\theta_0 = 120°$ (dashed line with asterisks), $\theta_0 = 150°$ (dotted line with squares), and $\theta_0 = 180°$ (dash-dot line with triangles). (a) DF and (b) WNG.

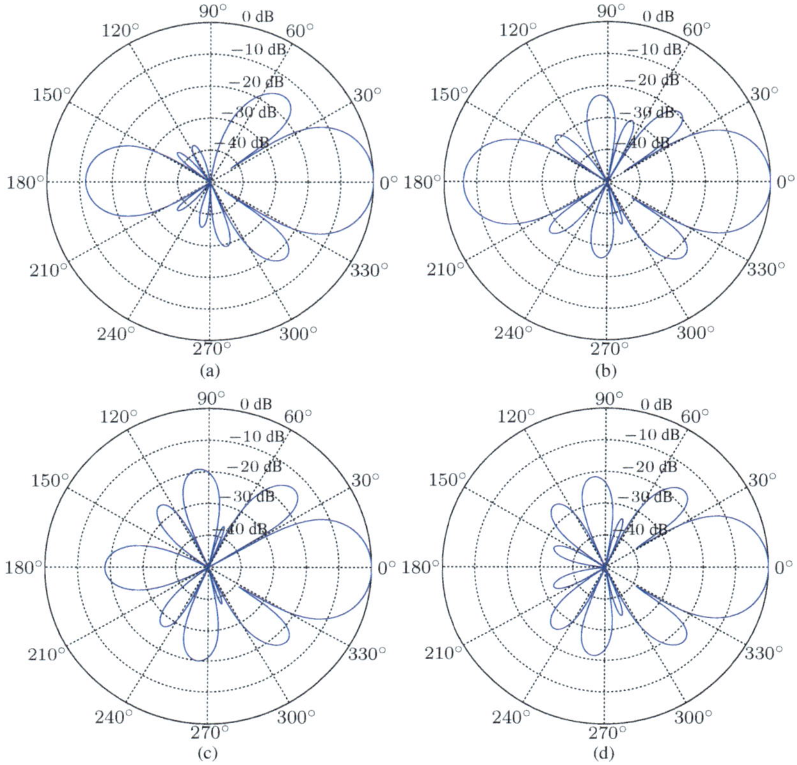

Fig. 6.12 Beampatterns of the global NS beamformer, $\mathbf{h}_{y,\mathrm{NS}} \otimes \mathbf{h}_{x,\mathrm{NS}}$, for $f = 2$ kHz, $\delta_x = \delta_y = 1$ cm, $M_x = 3$, $M_y = 5$, and several values of θ_0: (a) $\theta_0 = 90°$, (b) $\theta_0 = 120°$, (c) $\theta_0 = 150°$, and (d) $\theta_0 = 180°$.

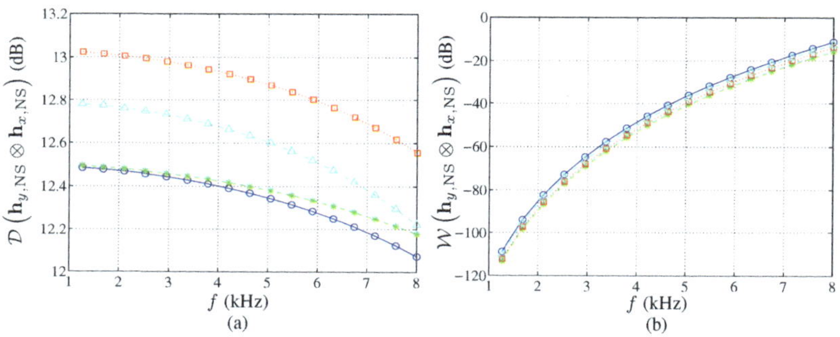

Fig. 6.13 Performance of the global NS beamformer, $\mathbf{h}_{y,\mathrm{NS}} \otimes \mathbf{h}_{x,\mathrm{NS}}$, as a function of frequency for $\delta_x = \delta_y = 1$ cm, $M_x = 3$, $M_y = 5$, and several values of θ_0: $\theta_0 = 90°$ (solid line with circles), $\theta_0 = 120°$ (dashed line with asterisks), $\theta_0 = 150°$ (dotted line with squares), and $\theta_0 = 180°$ (dash-dot line with triangles). (a) DF and (b) WNG.

References

1. H. L. Van Trees, *Optimum Array Processing: Part IV of Detection, Estimation, and Modulation Theory*. New York, NY: John Wiley & Sons, Inc., 2002.
2. J. Benesty, I. Cohen, and J. Chen, *Fundamentals of Signal Enhancement and Array Signal Processing*. Singapore: Wiley-IEEE Press, 2018.
3. H. Cox, R. M. Zeskind, and T. Kooij, "Practical supergain," *IEEE Trans. Acoust., Speech, Signal Process.*, vol. ASSP-34, pp. 393–398, June 1986.
4. H. Cox, R. M. Zeskind, and M. M. Owen, "Robust adaptive beamforming," *IEEE Trans. Acoust., Speech, Signal Process.*, vol. ASSP-35, pp. 1365–1376, Oct. 1987.
5. J. Benesty, J. Chen, and C. Pan, *Fundamentals of Differential Beamforming*. Springer Briefs in Electrical & Computer Engineering, 2016.

Chapter 7
Spatiotemporal Signal Enhancement

While previous chapters were mostly about (fixed, adaptive, and differential) beamforming with some specific array geometries, the discussion in this chapter is on the spatiotemporal signal enhancement problem with any array geometry. By taking into account the interframe correlation, we show how the Kronecker product appears naturally in the definition of the signal vector. Thanks to this, we explain how to perform noise reduction (i.e., signal enhancement) with Kronecker product filters and derive the most well-known algorithms.

7.1 Signal Model and Problem Formulation

We consider the conventional signal model in which an array of M sensors, with no specific geometry, captures a convolved source signal in some noise field. The received signals, at the time index t, are expressed as [1], [2], [3]

$$\begin{aligned} y_m(t) &= g_m(t) * s(t) + v_m(t) \\ &= x_m(t) + v_m(t), \ \ m = 1, 2, \ldots, M, \end{aligned} \tag{7.1}$$

where $g_m(t)$ is the impulse response from the unknown desired signal source, $s(t)$, location to the mth sensor, $*$ stands for linear convolution, and $v_m(t)$ is the additive noise at sensor m. We assume that the signals $x_m(t) = g_m(t)*s(t)$ and $v_m(t)$ are uncorrelated, zero mean, stationary, real, and broadband. By definition, the convolved desired signals, $x_m(t),\ m = 1, 2, \ldots, M$, are coherent across the array while the noise signals, $v_m(t),\ m = 1, 2, \ldots, M$, are typically only partially coherent across the array.

Using the short-time Fourier transform (STFT), (7.1) can be rewritten in the time-frequency domain as [4, 5]

J. Benesty et al., *Array Processing*, Springer Topics in Signal Processing 18, https://doi.org/10.1007/978-3-030-15600-8_7

$$\begin{aligned} Y_m(k,t') &= G_m(k)S(k,t') + V_m(k,t') \\ &= X_m(k,t') + V_m(k,t'),\ m = 1,2,\ldots,M, \end{aligned} \tag{7.2}$$

where $Y_m(k,t')$, $G_m(k)$, $S(k,t')$, $V_m(k,t')$, and $X_m(k,t')$ are the STFTs of $y_m(t)$, $g_m(t)$, $s(t)$, $v_m(t)$, and $x_m(t)$, respectively, at the frequency bin $k \in \{0,1,\ldots,K-1\}$ and the time frame t'. Assuming that the first sensor is the reference, we can write the M STFT-domain signals in a vector notation as

$$\begin{aligned} \mathbf{y}(k,t') &= \begin{bmatrix} Y_1(k,t') \ Y_2(k,t') \ \cdots \ Y_M(k,t') \end{bmatrix}^T \\ &= \mathbf{d}(k)X_1(k,t') + \mathbf{v}(k,t') \\ &= \mathbf{x}(k,t') + \mathbf{v}(k,t'), \end{aligned} \tag{7.3}$$

where

$$\mathbf{d}(k) = \begin{bmatrix} 1 \ \frac{G_2(k)}{G_1(k)} \ \cdots \ \frac{G_M(k)}{G_1(k)} \end{bmatrix}^T, \tag{7.4}$$

and $\mathbf{x}(k,t')$ and $\mathbf{v}(k,t')$ are defined similarly to $\mathbf{y}(k,t')$.

We assume that in the STFT domain, consecutive data frames are correlated and this information should be taken into account in order to improve the estimation of the desired signal. Considering the most recent L time frames, we can extend (7.3) to

$$\begin{aligned} \underline{\mathbf{y}}(k,t') &= \begin{bmatrix} \mathbf{y}^T(k,t') \ \mathbf{y}^T(k,t'-1) \ \cdots \ \mathbf{y}^T(k,t'-L+1) \end{bmatrix}^T \\ &= \mathbf{x}_1(k,t') \otimes \mathbf{d}(k) + \underline{\mathbf{v}}(k,t') \\ &= \underline{\mathbf{x}}(k,t') + \underline{\mathbf{v}}(k,t'), \end{aligned} \tag{7.5}$$

where

$$\mathbf{x}_1(k,t') = \begin{bmatrix} X_1(k,t') \ X_1(k,t'-1) \ \cdots \ X_1(k,t'-L+1) \end{bmatrix}^T \tag{7.6}$$

is a vector of length L, and $\underline{\mathbf{x}}(k,t')$ and $\underline{\mathbf{v}}(k,t')$ are vectors of length ML, defined similarly to $\underline{\mathbf{y}}(k,t')$. Then, the objective of spatiotemporal signal enhancement in the STFT domain (where the interframe correlation is taken into account) is to estimate $X_1(k,t')$ from $\underline{\mathbf{y}}(k,t')$ in the best possible way.

As one can observe, at the time frame t', our desired signal is $X_1(k,t')$ [and not the whole vector $\mathbf{x}_1(k,t')$]. However, $\mathbf{x}_1(k,t')$ contains both the desired signal, $X_1(k,t')$, and the components $X_1(k,t'-l)$, $l \neq 0$, which are not the desired signals at the time frame t' but signals that are correlated with $X_1(k,t')$. Therefore, the elements $X_1(k,t'-l)$, $l \neq 0$, contain both a part of the desired signal and a component that we consider as an interference. This suggests that we should decompose $X_1(k,t'-l)$ into two orthogonal components corresponding to the part of the desired signal and interference, i.e.,

$$X_1(k,t'-l) = \rho(k,t',l)X_1(k,t') + X_{1,\mathrm{i}}(k,t'-l), \tag{7.7}$$

where

$$X_{1,\mathrm{i}}(k,t'-l) = X_1(k,t'-l) - \rho(k,t',l)X_1(k,t'), \tag{7.8}$$

$$E\left[X_1^*(k,t')X_{1,\mathrm{i}}(k,t'-l)\right] = 0, \tag{7.9}$$

and

$$\rho(k,t',l) = \frac{E\left[X_1^*(k,t')X_1(k,t'-l)\right]}{E\left[|X_1(k,t')|^2\right]} \tag{7.10}$$

is the interframe correlation coefficient of the signal $X_1(k,t')$. Hence, we can write the vector $\underline{\mathbf{x}}(k,t')$ as

$$\begin{aligned}\underline{\mathbf{x}}(k,t') &= X_1(k,t')\left[\boldsymbol{\rho}(k,t') \otimes \mathbf{d}(k)\right] + \underline{\mathbf{x}}_{\mathrm{i}}(k,t') \\ &= X_1(k,t')\underline{\mathbf{d}}(k,t') + \underline{\mathbf{x}}_{\mathrm{i}}(k,t') \\ &= \underline{\mathbf{x}}_{\mathrm{d}}(k,t') + \underline{\mathbf{x}}_{\mathrm{i}}(k,t'),\end{aligned} \tag{7.11}$$

where

$$\boldsymbol{\rho}(k,t') = \begin{bmatrix} 1 \ \rho(k,t',1) \ \cdots \ \rho(k,t',L-1) \end{bmatrix}^T \tag{7.12}$$

is the (normalized) interframe correlation vector between $X_1(k,t')$ and $\mathbf{x}_1(k,t')$,

$$\underline{\mathbf{x}}_{\mathrm{i}}(k,t') = \begin{bmatrix} X_{1,\mathrm{i}}(k,t') \ X_{1,\mathrm{i}}(k,t'-1) \ \cdots \ X_{1,\mathrm{i}}(k,t'-L+1) \end{bmatrix}^T \otimes \mathbf{d}(k) \tag{7.13}$$

is the interference signal vector of length ML,

$$\underline{\mathbf{d}}(k,t') = \boldsymbol{\rho}(k,t') \otimes \mathbf{d}(k) \tag{7.14}$$

is a vector of length ML, and

$$\underline{\mathbf{x}}_{\mathrm{d}}(k,t') = X_1(k,t')\underline{\mathbf{d}}(k,t') \tag{7.15}$$

is the desired signal vector. As a consequence, we can express (7.5) as

$$\underline{\mathbf{y}}(k,t') = X_1(k,t')\left[\boldsymbol{\rho}(k,t') \otimes \mathbf{d}(k)\right] + \underline{\mathbf{x}}_{\mathrm{i}}(k,t') + \underline{\mathbf{v}}(k,t'). \tag{7.16}$$

In the rest, in order to simplify the notation, we often drop the dependence on k and t', so that (7.16), for example, is written as $\underline{\mathbf{y}} = X_1(\boldsymbol{\rho} \otimes \mathbf{d}) + \underline{\mathbf{x}}_{\mathrm{i}} + \underline{\mathbf{v}}$.

Since the desired, interference, and noise signals are mutually uncorrelated, the covariance matrix of $\underline{\mathbf{y}}$ is

$$\begin{aligned}\mathbf{\Phi}_{\underline{\mathbf{y}}} &= E\left(\underline{\mathbf{y}}\,\underline{\mathbf{y}}^H\right) \\ &= \mathbf{\Phi}_{\underline{\mathbf{x}}_{\mathrm{d}}} + \mathbf{\Phi}_{\underline{\mathbf{x}}_{\mathrm{i}}} + \mathbf{\Phi}_{\underline{\mathbf{v}}} \\ &= \mathbf{\Phi}_{\underline{\mathbf{x}}_{\mathrm{d}}} + \mathbf{\Phi}_{\mathrm{in}},\end{aligned} \tag{7.17}$$

where

$$\begin{aligned}\mathbf{\Phi}_{\underline{\mathbf{x}}_{\mathrm{d}}} &= \phi_{X_1}\underline{\mathbf{d}}\,\underline{\mathbf{d}}^H \\ &= \phi_{X_1}\left(\boldsymbol{\rho}\otimes\mathbf{d}\right)\left(\boldsymbol{\rho}\otimes\mathbf{d}\right)^H \\ &= \phi_{X_1}\left(\boldsymbol{\rho}\boldsymbol{\rho}^H\right)\otimes\left(\mathbf{d}\mathbf{d}^H\right)\end{aligned} \tag{7.18}$$

is the covariance matrix of $\underline{\mathbf{x}}_{\mathrm{d}}$, with $\phi_{X_1} = E\left(|X_1|^2\right)$ being the variance of X_1, $\mathbf{\Phi}_{\underline{\mathbf{v}}} = E\left(\underline{\mathbf{v}}\,\underline{\mathbf{v}}^H\right)$ is the covariance matrix of $\underline{\mathbf{v}}$, and $\mathbf{\Phi}_{\mathrm{in}} = \mathbf{\Phi}_{\underline{\mathbf{x}}_{\mathrm{i}}} + \mathbf{\Phi}_{\underline{\mathbf{v}}}$ is the covariance matrix of the interference-plus-noise $(\underline{\mathbf{x}}_{\mathrm{i}} + \underline{\mathbf{v}})$.

7.2 Signal Enhancement with Kronecker Product Filters

Given the structure of the desired signal vector, $\underline{\mathbf{x}}_{\mathrm{d}}$, we propose, as we did in previous chapters, to use complex-valued filters of length ML of the form:

$$\underline{\mathbf{h}} = \mathbf{h}_1 \otimes \mathbf{h}_2, \tag{7.19}$$

where $\mathbf{h}_1$ and $\mathbf{h}_2$ are two complex-valued linear filters of lengths L and M, respectively. Then, spatiotemporal signal enhancement is performed by applying $\underline{\mathbf{h}}$ to $\underline{\mathbf{y}}$, i.e.,

$$\begin{aligned}Z &= \underline{\mathbf{h}}^H\underline{\mathbf{y}} \\ &= \underline{\mathbf{h}}^H\underline{\mathbf{x}}_{\mathrm{d}} + \underline{\mathbf{h}}^H\underline{\mathbf{x}}_{\mathrm{i}} + \underline{\mathbf{h}}^H\underline{\mathbf{v}} \\ &= X_{\mathrm{fd}} + X_{\mathrm{ri}} + V_{\mathrm{rn}},\end{aligned} \tag{7.20}$$

where Z is the estimate of the desired signal, X_1,

$$\begin{aligned}X_{\mathrm{fd}} &= \left(\mathbf{h}_1\otimes\mathbf{h}_2\right)^H\left(\boldsymbol{\rho}\otimes\mathbf{d}\right)X_1 \\ &= \left(\mathbf{h}_1^H\boldsymbol{\rho}\right)\left(\mathbf{h}_2^H\mathbf{d}\right)X_1\end{aligned} \tag{7.21}$$

is the filtered desired signal,

$$X_{\mathrm{ri}} = \left(\mathbf{h}_1\otimes\mathbf{h}_2\right)^H\underline{\mathbf{x}}_{\mathrm{i}} \tag{7.22}$$

is the residual interference, and

$$V_{\mathrm{rn}} = \left(\mathbf{h}_1\otimes\mathbf{h}_2\right)^H\underline{\mathbf{v}} \tag{7.23}$$

is the residual noise. We deduce that the variance of Z is

$$\phi_Z = \phi_{X_{\rm fd}} + \phi_{X_{\rm ri}} + \phi_{V_{\rm rn}}, \tag{7.24}$$

where

$$\phi_{X_{\rm fd}} = \phi_{X_1} \left| \mathbf{h}_1^H \boldsymbol{\rho} \right|^2 \left| \mathbf{h}_2^H \mathbf{d} \right|^2, \tag{7.25}$$

$$\phi_{X_{\rm ri}} = \left(\mathbf{h}_1 \otimes \mathbf{h}_2 \right)^H \boldsymbol{\Phi}_{\underline{\mathbf{x}}_{\rm i}} \left(\mathbf{h}_1 \otimes \mathbf{h}_2 \right), \tag{7.26}$$

$$\phi_{V_{\rm rn}} = \left(\mathbf{h}_1 \otimes \mathbf{h}_2 \right)^H \boldsymbol{\Phi}_{\underline{\mathbf{v}}} \left(\mathbf{h}_1 \otimes \mathbf{h}_2 \right). \tag{7.27}$$

To end this section, we see that the distortionless constraints in our context are

$$\mathbf{h}_1^H \boldsymbol{\rho} = \mathbf{h}_2^H \mathbf{d} = 1. \tag{7.28}$$

7.3 Performance Measures

In the time-frequency domain, we must differentiate between the subband (i.e., single frequency) measures and the broadband (i.e., across the entire frequency range) measures. In this part, we define the most useful ones from the signal enhancement perspective.

We define the subband input SNR as

$$\mathrm{iSNR}(k, t') = \frac{\phi_{X_1}(k, t')}{\phi_{V_1}(k, t')}, \tag{7.29}$$

where $\phi_{V_1}(k, t') = E\left[\left| V_1(k, t') \right|^2 \right]$ is the variance of the noise at the first microphone. The broadband input SNR is obtained by summing over all time-frequency indices the numerator and denominator of $\mathrm{iSNR}(k, t')$. We get

$$\mathrm{iSNR} = \frac{\sum_{k,t'} \phi_{X_1}(k, t')}{\sum_{k,t'} \phi_{V_1}(k, t')}. \tag{7.30}$$

To quantify the level of the noise remaining at the output of the filter, we introduce the subband output SNR:

$$\begin{aligned} \mathrm{oSNR}\left[\underline{\mathbf{h}}(k, t') \right] &= \frac{\phi_{X_{\rm fd}}(k, t')}{\phi_{X_{\rm ri}}(k, t') + \phi_{V_{\rm rn}}(k, t')} \\ &= \frac{\phi_{X_1}(k, t') \left| \underline{\mathbf{h}}^H(k, t') \underline{\mathbf{d}}(k, t') \right|^2}{\underline{\mathbf{h}}^H(k, t') \boldsymbol{\Phi}_{\rm in}(k, t') \underline{\mathbf{h}}(k, t')} \end{aligned} \tag{7.31}$$

and the broadband output SNR:

$$\mathrm{oSNR}\left(\underline{\mathbf{h}}\right) = \frac{\sum_{k,t'} \phi_{X_1}(k,t') \left| \underline{\mathbf{h}}^H(k,t')\underline{\mathbf{d}}(k,t') \right|^2}{\sum_{k,t'} \underline{\mathbf{h}}^H(k,t')\mathbf{\Phi}_{\mathrm{in}}(k,t')\underline{\mathbf{h}}(k,t')}. \tag{7.32}$$

We define the subband and broadband array gains as

$$\mathcal{G}\left[\underline{\mathbf{h}}(k,t')\right] = \frac{\mathrm{oSNR}\left[\underline{\mathbf{h}}(k,t')\right]}{\mathrm{iSNR}(k,t')} \tag{7.33}$$

and

$$\mathcal{G}\left(\underline{\mathbf{h}}\right) = \frac{\mathrm{oSNR}\left(\underline{\mathbf{h}}\right)}{\mathrm{iSNR}}. \tag{7.34}$$

It is easy to see that the maximum subband array gain is

$$\mathcal{G}_{\max} = \phi_{V_1} \underline{\mathbf{d}}^H \mathbf{\Phi}_{\mathrm{in}}^{-1} \underline{\mathbf{d}} \geq 1. \tag{7.35}$$

The noise reduction factor quantifies the amount of the noise that is rejected by the filter. The subband noise reduction factor is then

$$\begin{aligned} \xi_{\mathrm{n}}\left[\underline{\mathbf{h}}(k,t')\right] &= \frac{\phi_{V_1}(k,t')}{\phi_{X_{\mathrm{ri}}}(k,t') + \phi_{V_{\mathrm{rn}}}(k,t')} \\ &= \frac{\phi_{V_1}(k,t')}{\underline{\mathbf{h}}^H(k,t')\mathbf{\Phi}_{\mathrm{in}}(k,t')\underline{\mathbf{h}}(k,t')} \end{aligned} \tag{7.36}$$

and the broadband noise reduction factor is

$$\xi_{\mathrm{n}}\left(\underline{\mathbf{h}}\right) = \frac{\sum_{k,t'} \phi_{V_1}(k,t')}{\sum_{k,t'} \underline{\mathbf{h}}^H(k,t')\mathbf{\Phi}_{\mathrm{in}}(k,t')\underline{\mathbf{h}}(k,t')}. \tag{7.37}$$

The noise reduction factor is expected to be lower bounded by 1 for optimal filters; so the more the noise is reduced, the higher its value.

The desired signal can be distorted by the filter. Therefore, we define the subband desired signal reduction factor as

$$\begin{aligned} \xi_{\mathrm{d}}\left[\underline{\mathbf{h}}(k,t')\right] &= \frac{\phi_{X_1}(k,t')}{\phi_{X_{\mathrm{fd}}}(k,t')} \\ &= \frac{1}{\left| \underline{\mathbf{h}}^H(k,t')\underline{\mathbf{d}}(k,t') \right|^2} \end{aligned} \tag{7.38}$$

and the broadband desired signal reduction factor as

$$\xi_{\mathrm{d}}\left(\underline{\mathbf{h}}\right) = \frac{\sum_{k,t'} \phi_{X_1}(k,t')}{\sum_{k,t'} \phi_{X_1}(k,t') \left| \underline{\mathbf{h}}^H(k,t')\underline{\mathbf{d}}(k,t') \right|^2}. \tag{7.39}$$

An important observation is that the design of a filter that does not distort the desired signal requires the constraint:

$$\underline{\mathbf{h}}^H \underline{\mathbf{d}} = 1. \tag{7.40}$$

Thus, the desired signal reduction factor is equal to 1 if there is no distortion and expected to be greater than 1 when distortion occurs.

It is easy to verify the fundamental relation:

$$\frac{\mathrm{oSNR}\left(\underline{\mathbf{h}}\right)}{\mathrm{iSNR}} = \frac{\xi_{\mathrm{n}}\left(\underline{\mathbf{h}}\right)}{\xi_{\mathrm{d}}\left(\underline{\mathbf{h}}\right)}. \tag{7.41}$$

Another useful performance measure is the subband desired signal distortion index given by

$$\begin{aligned} \upsilon_{\mathrm{d}}\left[\underline{\mathbf{h}}(k,t')\right] &= \frac{E\left[\left|X_{\mathrm{fd}}(k,t') - X_1(k,t')\right|^2\right]}{\phi_{X_1}(k,t')} \\ &= \left| \underline{\mathbf{h}}^H(k,t')\underline{\mathbf{d}}(k,t') - 1 \right|^2. \end{aligned} \tag{7.42}$$

The broadband desired signal distortion index is given by

$$\upsilon_{\mathrm{d}}\left(\underline{\mathbf{h}}\right) = \frac{\sum_{k,t'} \phi_{X_1}(k,t') \left| \underline{\mathbf{h}}^H(k,t')\underline{\mathbf{d}}(k,t') - 1 \right|^2}{\sum_{k,t'} \phi_{X_1}(k,t')}. \tag{7.43}$$

The desired signal distortion index is always greater than or equal to 0 and should be upper bounded by 1 for optimal filters; so the higher is its value, the more distortion to the desired signal.

Now, let us write the error signal between the estimated and desired signals:

$$\begin{aligned} \mathcal{E}(k,t') &= Z(k,t') - X_1(k,t') \\ &= X_{\mathrm{fd}}(k,t') + X_{\mathrm{ri}}(k,t') + V_{\mathrm{rn}}(k,t') - X_1(k,t') \\ &= \mathcal{E}_{\mathrm{d}}(k,t') + \mathcal{E}_{\mathrm{r}}(k,t'), \end{aligned} \tag{7.44}$$

where

$$\begin{aligned} \mathcal{E}_{\mathrm{d}}(k,t') &= X_{\mathrm{fd}}(k,t') - X_1(k,t') \\ &= \left[\underline{\mathbf{h}}^H(k,t')\underline{\mathbf{d}}(k,t') - 1 \right] X_1(k,t') \end{aligned} \tag{7.45}$$

is the desired signal distortion due to the complex filter and

$$\begin{aligned}\mathcal{E}_{\mathrm{r}}(k,t') &= X_{\mathrm{ri}}(k,t') + V_{\mathrm{rn}}(k,t') \\ &= \underline{\mathbf{h}}^H(k,t')\underline{\mathbf{x}}_{\mathrm{i}}(k,t') + \underline{\mathbf{h}}^H(k,t')\underline{\mathbf{v}}(k,t') \end{aligned} \tag{7.46}$$

represents the residual interference-plus-noise.

Having defined the error signal, we can write the subband MSE criterion:

$$\begin{aligned} J\left[\underline{\mathbf{h}}(k,t')\right] &= E\left[\left|\mathcal{E}(k,t')\right|^2\right] \\ &= J_{\mathrm{d}}\left[\underline{\mathbf{h}}(k,t')\right] + J_{\mathrm{n}}\left[\underline{\mathbf{h}}(k,t')\right], \end{aligned} \tag{7.47}$$

where

$$\begin{aligned} J_{\mathrm{d}}\left[\underline{\mathbf{h}}(k,t')\right] &= E\left[\left|\mathcal{E}_{\mathrm{d}}(k,t')\right|^2\right] \\ &= E\left[\left|X_{\mathrm{fd}}(k,t') - X_1(k,t')\right|^2\right] \\ &= \phi_{X_1}(k,t')\left|\underline{\mathbf{h}}^H(k,t')\underline{\mathbf{d}}(k,t') - 1\right|^2 \\ &= \phi_{X_1}(k,t') \times \upsilon_{\mathrm{d}}\left[\underline{\mathbf{h}}(k,t')\right] \end{aligned} \tag{7.48}$$

and

$$\begin{aligned} J_{\mathrm{n}}\left[\underline{\mathbf{h}}(k,t')\right] &= E\left[\left|\mathcal{E}_{\mathrm{r}}(k,t')\right|^2\right] \\ &= E\left[\left|X_{\mathrm{ri}}(k,t')\right|^2\right] + E\left[\left|V_{\mathrm{rn}}(k,t')\right|^2\right] \\ &= \phi_{X_{\mathrm{ri}}}(k,t') + \phi_{V_{\mathrm{rn}}}(k,t') \\ &= \frac{\phi_{V_1}(k,t')}{\xi_{\mathrm{n}}\left[\underline{\mathbf{h}}(k,t')\right]}. \end{aligned} \tag{7.49}$$

The broadband MSE is given by

$$\begin{aligned} J\left(\underline{\mathbf{h}}\right) &= \sum_{k,t'} J\left[\underline{\mathbf{h}}(k,t')\right] \qquad (7.50) \\ &= \sum_{k,t'} J_{\mathrm{d}}\left[\underline{\mathbf{h}}(k,t')\right] + \sum_{k,t'} J_{\mathrm{n}}\left[\underline{\mathbf{h}}(k,t')\right] \\ &= J_{\mathrm{d}}\left(\underline{\mathbf{h}}\right) + J_{\mathrm{n}}\left(\underline{\mathbf{h}}\right). \end{aligned}$$

7.4 Optimal Signal Enhancement Kronecker Product Filters

In this section, we present three signal enhancement algorithms based on Kronecker product filters. Of course, many more can be derived but we prefer to focus on the most obvious ones only.

7.4.1 Wiener

Because of the structure of $\underline{\mathbf{h}}$, it does not seem possible to derive a closed-form Wiener filter but we can write the subband MSE in (7.47) as

$$J(\mathbf{h}_1 \otimes \mathbf{h}_2) = \phi_{X_1} + \mathbf{h}_1^H \mathbf{\Phi}_{\underline{\mathbf{y}},2} \mathbf{h}_1 - \phi_{X_1,2} \mathbf{h}_1^H \boldsymbol{\rho} - \phi_{X_1,2}^* \boldsymbol{\rho}^H \mathbf{h}_1 \tag{7.51}$$

$$= \phi_{X_1} + \mathbf{h}_2^H \mathbf{\Phi}_{\underline{\mathbf{y}},1} \mathbf{h}_2 - \phi_{X_1,1} \mathbf{h}_2^H \mathbf{d} - \phi_{X_1,1}^* \mathbf{d}^H \mathbf{h}_2, \tag{7.52}$$

where

$$\mathbf{\Phi}_{\underline{\mathbf{y}},2} = (\mathbf{I}_L \otimes \mathbf{h}_2)^H \mathbf{\Phi}_{\underline{\mathbf{y}}} (\mathbf{I}_L \otimes \mathbf{h}_2), \tag{7.53}$$

$$\phi_{X_1,2} = \phi_{X_1} \mathbf{h}_2^H \mathbf{d}, \tag{7.54}$$

and

$$\mathbf{\Phi}_{\underline{\mathbf{y}},1} = (\mathbf{h}_1 \otimes \mathbf{I}_M)^H \mathbf{\Phi}_{\underline{\mathbf{y}}} (\mathbf{h}_1 \otimes \mathbf{I}_M), \tag{7.55}$$

$$\phi_{X_1,1} = \phi_{X_1} \mathbf{h}_1^H \boldsymbol{\rho}. \tag{7.56}$$

It is important to observe that the sizes of the matrices $\mathbf{\Phi}_{\underline{\mathbf{y}},1}$ and $\mathbf{\Phi}_{\underline{\mathbf{y}},2}$, which are $M \times M$ and $L \times L$, respectively, are much smaller than the size of $\mathbf{\Phi}_{\underline{\mathbf{y}}}$, which is $ML \times ML$. As a result, in practice, much less observations are needed to accurately estimate $\mathbf{\Phi}_{\underline{\mathbf{y}},1}$ and $\mathbf{\Phi}_{\underline{\mathbf{y}},2}$ than $\mathbf{\Phi}_{\underline{\mathbf{y}}}$, which is the matrix that is inverted in the multichannel Wiener filter taking into account the interframe correlation.

When $\mathbf{h}_2$ is fixed, we write (7.51) as

$$J(\mathbf{h}_1 | \mathbf{h}_2) = \phi_{X_1} + \mathbf{h}_1^H \mathbf{\Phi}_{\underline{\mathbf{y}},2} \mathbf{h}_1 - \phi_{X_1,2} \mathbf{h}_1^H \boldsymbol{\rho} - \phi_{X_1,2}^* \boldsymbol{\rho}^H \mathbf{h}_1, \tag{7.57}$$

and when $\mathbf{h}_1$ is fixed, we write (7.52) as

$$J(\mathbf{h}_2 | \mathbf{h}_1) = \phi_{X_1} + \mathbf{h}_2^H \mathbf{\Phi}_{\underline{\mathbf{y}},1} \mathbf{h}_2 - \phi_{X_1,1} \mathbf{h}_2^H \mathbf{d} - \phi_{X_1,1}^* \mathbf{d}^H \mathbf{h}_2. \tag{7.58}$$

Now, we have everything to derive an iterative algorithm similar to the one proposed in [6]. At iteration 0, we may take

$$\mathbf{h}_2^{(0)} = \phi_{X_1} \mathbf{\Phi}_{\mathbf{y}}^{-1} \mathbf{d}, \tag{7.59}$$

where $\mathbf{\Phi}_{\mathbf{y}} = E\left(\mathbf{y}\mathbf{y}^H\right)$ is the covariance matrix of $\mathbf{y}$ [see (7.3)]. In fact, $\mathbf{h}_2^{(0)}$ is the conventional multichannel Wiener filter, which does not take into account the interframe correlation. Substituting $\mathbf{h}_2^{(0)}$ into (7.53)–(7.54), we get

$$\mathbf{\Phi}_{\underline{\mathbf{y}},2}^{(0)} = \left(\mathbf{I}_L \otimes \mathbf{h}_2^{(0)}\right)^H \mathbf{\Phi}_{\underline{\mathbf{y}}} \left(\mathbf{I}_L \otimes \mathbf{h}_2^{(0)}\right), \tag{7.60}$$

$$\phi_{X_1,2}^{(0)} = \phi_{X_1} \left(\mathbf{h}_2^{(0)}\right)^H \mathbf{d}. \tag{7.61}$$

Then, substituting these quantities into the MSE in (7.57), we obtain at iteration 1:

$$J\left(\mathbf{h}_1^{(1)}|\mathbf{h}_2^{(0)}\right) = \phi_{X_1} + \left(\mathbf{h}_1^{(1)}\right)^H \mathbf{\Phi}_{\underline{\mathbf{y}},2}^{(0)}\mathbf{h}_1^{(1)} - \phi_{X_1,2}^{(0)}\left(\mathbf{h}_1^{(1)}\right)^H \boldsymbol{\rho} - \left(\phi_{X_1,2}^{(0)}\right)^* \boldsymbol{\rho}^H\mathbf{h}_1^{(1)}. \tag{7.62}$$

The minimization of $J\left(\mathbf{h}_1^{(1)}|\mathbf{h}_2^{(0)}\right)$ with respect to $\mathbf{h}_1^{(1)}$ gives

$$\mathbf{h}_1^{(1)} = \phi_{X_1,2}^{(0)}\left(\mathbf{\Phi}_{\underline{\mathbf{y}},2}^{(0)}\right)^{-1}\boldsymbol{\rho}. \tag{7.63}$$

Using $\mathbf{h}_1^{(1)}$ into (7.55)–(7.56), we obtain

$$\mathbf{\Phi}_{\underline{\mathbf{y}},1}^{(1)} = \left(\mathbf{h}_1^{(1)} \otimes \mathbf{I}_M\right)^H \mathbf{\Phi}_{\underline{\mathbf{y}}}\left(\mathbf{h}_1^{(1)} \otimes \mathbf{I}_M\right), \tag{7.64}$$

$$\phi_{X_1,1}^{(1)} = \phi_{X_1}\left(\mathbf{h}_1^{(1)}\right)^H \boldsymbol{\rho}. \tag{7.65}$$

With $\mathbf{\Phi}_{\underline{\mathbf{y}},1}^{(1)}$ and $\phi_{X_1,1}^{(1)}$, we can compute the MSE in (7.58) as

$$J\left(\mathbf{h}_2^{(1)}|\mathbf{h}_1^{(1)}\right) = \phi_{X_1} + \left(\mathbf{h}_2^{(1)}\right)^H \mathbf{\Phi}_{\underline{\mathbf{y}},1}^{(1)}\mathbf{h}_2^{(1)} - \phi_{X_1,1}^{(1)}\left(\mathbf{h}_2^{(1)}\right)^H \mathbf{d} - \left(\phi_{X_1,1}^{(1)}\right)^* \mathbf{d}^H\mathbf{h}_2^{(1)}, \tag{7.66}$$

whose minimization with respect to $\mathbf{h}_2^{(1)}$ gives

$$\mathbf{h}_2^{(1)} = \phi_{X_1,1}^{(1)}\left(\mathbf{\Phi}_{\underline{\mathbf{y}},1}^{(1)}\right)^{-1}\mathbf{d}. \tag{7.67}$$

Continuing the iterations up to the iteration n, we easily get the estimate of the first filter:

$$\mathbf{h}_1^{(n)} = \phi_{X_1,2}^{(n-1)}\left(\mathbf{\Phi}_{\underline{\mathbf{y}},2}^{(n-1)}\right)^{-1}\boldsymbol{\rho}, \tag{7.68}$$

where

$$\phi_{X_1,2}^{(n-1)} = \phi_{X_1}\left(\mathbf{h}_2^{(n-1)}\right)^H \mathbf{d}, \tag{7.69}$$

$$\mathbf{\Phi}_{\underline{\mathbf{y}},2}^{(n-1)} = \left(\mathbf{I}_L \otimes \mathbf{h}_2^{(n-1)}\right)^H \mathbf{\Phi}_{\underline{\mathbf{y}}}\left(\mathbf{I}_L \otimes \mathbf{h}_2^{(n-1)}\right), \tag{7.70}$$

and the estimate of the second filter:

$$\mathbf{h}_2^{(n)} = \phi_{X_1,1}^{(n)}\left(\mathbf{\Phi}_{\underline{\mathbf{y}},1}^{(n)}\right)^{-1}\mathbf{d}, \tag{7.71}$$

where

$$\phi_{X_1,1}^{(n)} = \phi_{X_1} \left(\mathbf{h}_1^{(n)}\right)^H \boldsymbol{\rho}, \quad (7.72)$$

$$\boldsymbol{\Phi}_{\underline{\mathbf{y}},1}^{(n)} = \left(\mathbf{h}_1^{(n)} \otimes \mathbf{I}_M\right)^H \boldsymbol{\Phi}_{\underline{\mathbf{y}}} \left(\mathbf{h}_1^{(n)} \otimes \mathbf{I}_M\right). \quad (7.73)$$

Finally, we deduce that the Wiener filter is at iteration n:

$$\underline{\mathbf{h}}_{\mathrm{W}}^{(n)} = \mathbf{h}_1^{(n)} \otimes \mathbf{h}_2^{(n)}, \quad (7.74)$$

where $\mathbf{h}_1^{(n)}$ and $\mathbf{h}_2^{(n)}$ are defined in (7.68) and (7.71), respectively.

Example 7.1. Consider a ULA of M sensors with an interelement spacing equal to δ. Suppose that a desired speech signal, $s(t)$, impinges on the ULA from the direction θ_s, and that an interference $u(t)$ impinges on the ULA from the direction θ_u. Assume that the interference $u(t)$ is white Gaussian noise, i.e., $u(t) \sim \mathcal{N}\left(0, \sigma_u^2\right)$, uncorrelated with $s(t)$. In addition, the sensors contain thermal white Gaussian noise, $w_m(t) \sim \mathcal{N}\left(0, \sigma_w^2\right)$, that are mutually uncorrelated. The desired speech signal, $X_1(k, t')$, needs to be recovered from the noisy received signals, $y_m(t) = x_m(t) + v_m(t)$, $m = 1, 2, \ldots, M$, where $v_m(t) = u_m(t) + w_m(t)$, $m = 1, 2, \ldots, M$ are the additive noise signals.

To demonstrate spatiotemporal speech enhancement in the STFT domain, we choose a sampling frequency of 16 kHz, $\delta = 2$ cm, $\theta_s = 0$, $\theta_u = 50°$, $\sigma_w^2 = 0.1\sigma_u^2$, a Hamming window of length $L = 512$ as the analysis window, overlap of 75% between consecutive windows (384 samples), and the Wiener filter at the iteration $n = 5$, $\underline{\mathbf{h}}_{\mathrm{W}}^{(5)}$, in the STFT domain.

Figure 7.1 shows the spectrogram and waveform of the clean speech signal received at the first sensor, $x_1(t)$. Figure 7.2 shows plots of the broadband gain in SNR, $\mathcal{G}\left(\underline{\mathbf{h}}_{\mathrm{W}}^{(5)}\right)$, the broadband MSE, $J\left(\underline{\mathbf{h}}_{\mathrm{W}}^{(5)}\right)$, the broadband noise reduction factor, $\xi_{\mathrm{n}}\left(\underline{\mathbf{h}}_{\mathrm{W}}^{(5)}\right)$, and the broadband desired signal reduction factor, $\xi_{\mathrm{d}}\left(\underline{\mathbf{h}}_{\mathrm{W}}^{(5)}\right)$, as a function of the broadband input SNR, for $L = 2$ and different numbers of sensors, M. Figure 7.3 shows a realization of the observation signal at the first sensor, $y_1(t)$, and the estimated signals, $z(t)$, for different numbers of time frames, L, and different numbers of sensors, M. Clearly, as the number of time frames or the number of sensors increases, the Wiener filter better enhances the desired speech signal in terms of higher SNR and noise reduction, and lower MSE and desired signal reduction.

7.4.2 Tradeoff

In order to better compromise between noise reduction and signal distortion, we propose to minimize the desired signal distortion indices with the con-

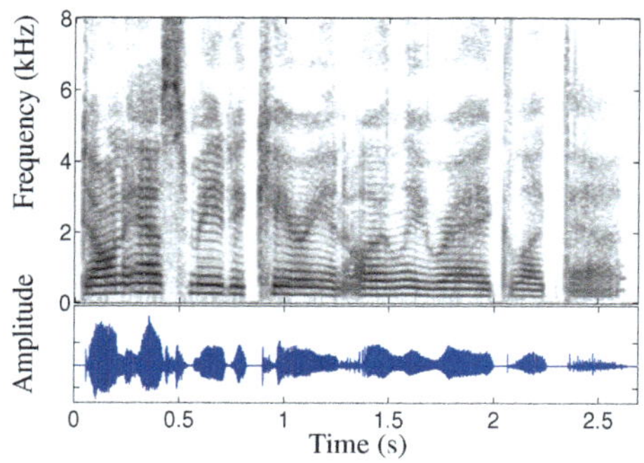

Fig. 7.1 Speech spectrogram and waveform of a clean speech signal received at the first sensor, $x_1(t)$: "Don't ask me to carry an oily rag like that."

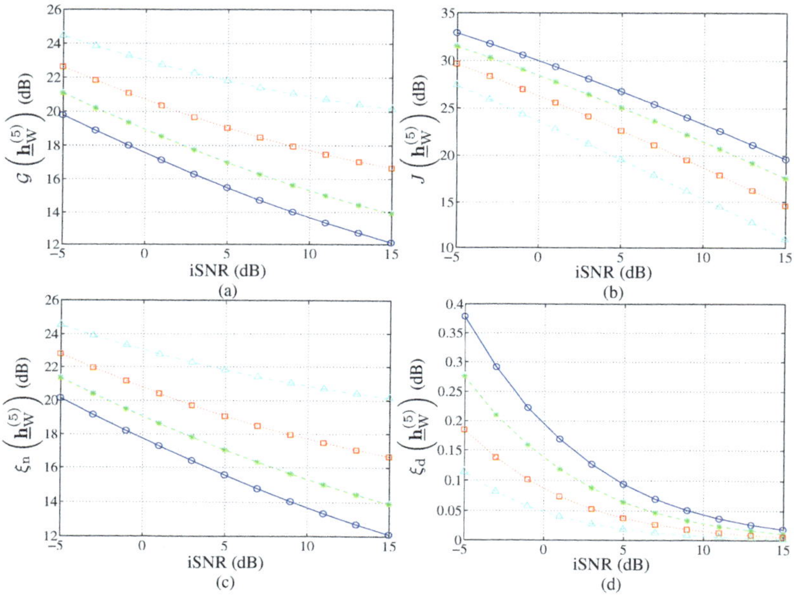

Fig. 7.2 (a) The broadband gain in SNR, (b) the broadband MSE, (c) the broadband noise reduction factor, and (d) the broadband desired signal reduction factor of the Wiener filter, $\underline{\mathbf{h}}_{\mathrm{W}}^{(5)}$, as a function of the broadband input SNR, for $L = 2$ and different numbers of sensors, M: $M = 1$ (solid line with circles), $M = 2$ (dashed line with asterisks), $M = 4$ (dotted line with squares), and $M = 8$ (dash-dot line with triangles).

straints that the noise reduction factors are equal to positive values that are greater than 1, i.e.,

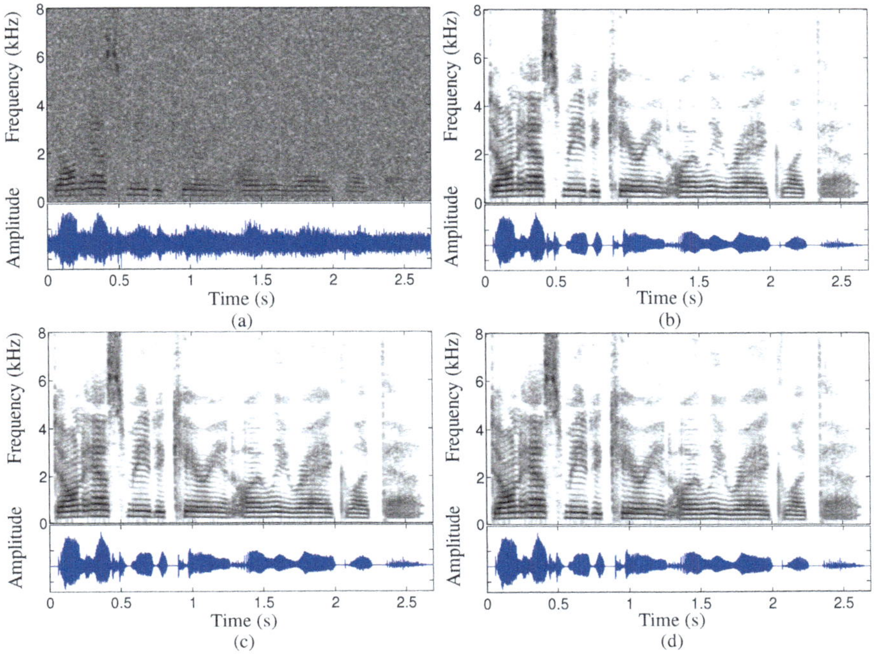

Fig. 7.3 Speech spectrograms and waveforms of (a) the noisy speech signal received at the first sensor, $Y_1(k,t')$ (iSNR = 0 dB), and the estimated signal, $Z(k,t')$, for (b) $M = L = 2$ [oSNR $\left(\underline{\mathbf{h}}_{\mathrm{W}}^{(5)}\right) = 18.91$ dB], (c) $M = 2$, $L = 4$ [oSNR $\left(\underline{\mathbf{h}}_{\mathrm{W}}^{(5)}\right) = 22.36$ dB], and (d) $M = L = 4$ [oSNR $\left(\underline{\mathbf{h}}_{\mathrm{W}}^{(5)}\right) = 24.01$ dB].

$$\min_{\mathbf{h}_1^{(n)}} J_{\mathrm{d}}\left(\mathbf{h}_1^{(n)}|\mathbf{h}_2^{(n-1)}\right) \quad \text{subject to} \quad J_{\mathrm{n}}\left(\mathbf{h}_1^{(n)}|\mathbf{h}_2^{(n-1)}\right) = \aleph_1 \phi_{V_1}, \tag{7.75}$$

$$\min_{\mathbf{h}_2^{(n)}} J_{\mathrm{d}}\left(\mathbf{h}_2^{(n)}|\mathbf{h}_1^{(n)}\right) \quad \text{subject to} \quad J_{\mathrm{n}}\left(\mathbf{h}_2^{(n)}|\mathbf{h}_1^{(n)}\right) = \aleph_2 \phi_{V_1}, \tag{7.76}$$

where $0 < \aleph_1, \aleph_2 < 1$ to insure that we get some noise reduction,

$$\begin{aligned} J_{\mathrm{d}}\left(\mathbf{h}_1^{(n)}|\mathbf{h}_2^{(n-1)}\right) &= \phi_{X_1} + \left(\mathbf{h}_1^{(n)}\right)^H \mathbf{\Phi}_{\underline{\mathbf{x}}_{\mathrm{d}},2}^{(n-1)} \mathbf{h}_1^{(n)} - \phi_{X_1,2}^{(n-1)} \left(\mathbf{h}_1^{(n)}\right)^H \boldsymbol{\rho} \\ &\quad - \left(\phi_{X_1,2}^{(n-1)}\right)^* \boldsymbol{\rho}^H \mathbf{h}_1^{(n)}, \end{aligned} \tag{7.77}$$

$$J_{\mathrm{n}}\left(\mathbf{h}_1^{(n)}|\mathbf{h}_2^{(n-1)}\right) = \left(\mathbf{h}_1^{(n)}\right)^H \mathbf{\Phi}_{\mathrm{in},2}^{(n-1)} \mathbf{h}_1^{(n)}, \tag{7.78}$$

and

$$J_{\mathrm{d}}\left(\mathbf{h}_2^{(n)}|\mathbf{h}_1^{(n)}\right) = \phi_{X_1} + \left(\mathbf{h}_2^{(n)}\right)^H \mathbf{\Phi}_{\underline{\mathbf{x}}_{\mathrm{d}},1}^{(n)}\mathbf{h}_2^{(n)} - \phi_{X_1,1}^{(n)}\left(\mathbf{h}_2^{(n)}\right)^H \mathbf{d} - \left(\phi_{X_1,1}^{(n)}\right)^* \mathbf{d}^H\mathbf{h}_2^{(n)}, \tag{7.79}$$

$$J_{\mathrm{n}}\left(\mathbf{h}_2^{(n)}|\mathbf{h}_1^{(n)}\right) = \left(\mathbf{h}_2^{(n)}\right)^H \mathbf{\Phi}_{\mathrm{in},1}^{(n)}\mathbf{h}_2^{(n)}, \tag{7.80}$$

with

$$\mathbf{\Phi}_{\underline{\mathbf{x}}_{\mathrm{d}},2}^{(n-1)} = \left(\mathbf{I}_L \otimes \mathbf{h}_2^{(n-1)}\right)^H \mathbf{\Phi}_{\underline{\mathbf{x}}_{\mathrm{d}}}\left(\mathbf{I}_L \otimes \mathbf{h}_2^{(n-1)}\right) = \frac{\left|\phi_{X_1,2}^{(n-1)}\right|^2}{\phi_{X_1}}\boldsymbol{\rho}\boldsymbol{\rho}^H, \tag{7.81}$$

$$\mathbf{\Phi}_{\mathrm{in},2}^{(n-1)} = \left(\mathbf{I}_L \otimes \mathbf{h}_2^{(n-1)}\right)^H \mathbf{\Phi}_{\mathrm{in}}\left(\mathbf{I}_L \otimes \mathbf{h}_2^{(n-1)}\right), \tag{7.82}$$

$$\mathbf{\Phi}_{\underline{\mathbf{x}}_{\mathrm{d}},1}^{(n)} = \left(\mathbf{h}_1^{(n)} \otimes \mathbf{I}_M\right)^H \mathbf{\Phi}_{\underline{\mathbf{x}}_{\mathrm{d}}}\left(\mathbf{h}_1^{(n)} \otimes \mathbf{I}_M\right) = \frac{\left|\phi_{X_1,1}^{(n)}\right|^2}{\phi_{X_1}}\mathbf{d}\mathbf{d}^H, \tag{7.83}$$

$$\mathbf{\Phi}_{\mathrm{in},1}^{(n)} = \left(\mathbf{h}_1^{(n)} \otimes \mathbf{I}_M\right)^H \mathbf{\Phi}_{\mathrm{in}}\left(\mathbf{h}_1^{(n)} \otimes \mathbf{I}_M\right). \tag{7.84}$$

By using Lagrange multipliers, $\mu_1, \mu_2 > 0$, to adjoin the constraints to the cost functions, we get

$$\mathbf{h}_{1,\mu_1}^{(n)} = \phi_{X_1,2}^{(n-1)}\left(\mathbf{\Phi}_{\underline{\mathbf{x}}_{\mathrm{d}},2}^{(n-1)} + \mu_1\mathbf{\Phi}_{\mathrm{in},2}^{(n-1)}\right)^{-1}\boldsymbol{\rho} = \frac{\phi_{X_1,2}^{(n-1)}\phi_{X_1}\left(\mathbf{\Phi}_{\mathrm{in},2}^{(n-1)}\right)^{-1}\boldsymbol{\rho}}{\mu_1\phi_{X_1} + \left|\phi_{X_1,2}^{(n-1)}\right|^2\boldsymbol{\rho}^H\left(\mathbf{\Phi}_{\mathrm{in},2}^{(n-1)}\right)^{-1}\boldsymbol{\rho}} \tag{7.85}$$

and

$$\mathbf{h}_{2,\mu_2}^{(n)} = \phi_{X_1,1}^{(n)}\left(\mathbf{\Phi}_{\underline{\mathbf{x}}_{\mathrm{d}},1}^{(n)} + \mu_2\mathbf{\Phi}_{\mathrm{in},1}^{(n)}\right)^{-1}\mathbf{d} = \frac{\phi_{X_1,1}^{(n)}\phi_{X_1}\left(\mathbf{\Phi}_{\mathrm{in},1}^{(n)}\right)^{-1}\mathbf{d}}{\mu_2\phi_{X_1} + \left|\phi_{X_1,1}^{(n)}\right|^2\mathbf{d}^H\left(\mathbf{\Phi}_{\mathrm{in},1}^{(n)}\right)^{-1}\mathbf{d}}, \tag{7.86}$$

with the initialization:

$$\mathbf{h}_{2,\mu_2}^{(0)} = \phi_{X_1}\left(\phi_{X_1}\mathbf{d}\mathbf{d}^H + \mu_2\mathbf{\Phi}_{\mathbf{v}}\right)^{-1}\mathbf{d} = \frac{\phi_{X_1}\mathbf{\Phi}_{\mathbf{v}}^{-1}\mathbf{d}}{\mu_2 + \phi_{X_1}\mathbf{d}^H\mathbf{\Phi}_{\mathbf{v}}^{-1}\mathbf{d}}, \tag{7.87}$$

where $\mathbf{\Phi}_{\mathbf{v}} = E\left(\mathbf{v}\mathbf{v}^H\right)$ is the covariance matrix of $\mathbf{v}$ [see (7.3)]. As a matter of fact, $\mathbf{h}_{2,\mu_2}^{(0)}$ is just the traditional multichannel tradeoff filter. Therefore, we find that the tradeoff filter is at iteration n:

$$\underline{\mathbf{h}}_{\mathrm{T},\mu_1,\mu_2}^{(n)} = \mathbf{h}_{1,\mu_1}^{(n)} \otimes \mathbf{h}_{2,\mu_2}^{(n)}, \tag{7.88}$$

where $\mathbf{h}_{1,\mu_1}^{(n)}$ and $\mathbf{h}_{2,\mu_2}^{(n)}$ are defined in (7.85) and (7.86), respectively. We can see that for

- $\mu_1 = \mu_2 = 1$, we get the Wiener filter;
- $\mu_1, \mu_2 > 1$, results in a filter with low residual noise at the expense of high signal distortion (as compared to Wiener); and
- $\mu_1, \mu_2 < 1$, results in a filter with high residual noise and low signal distortion (as compared to Wiener).

Example 7.2. Returning to Example 7.1, we now employ the tradeoff filter at the iteration $n = 5$, $\underline{\mathbf{h}}_{\mathrm{T},\mu_1,\mu_2}^{(5)}$. We assume $L = 2$ time frames and $M = 4$ sensors. Figure 7.4 shows plots of the broadband gain in SNR, $\mathcal{G}\left(\underline{\mathbf{h}}_{\mathrm{T},\mu_1,\mu_2}^{(5)}\right)$, the broadband desired signal distortion index, $\upsilon_{\mathrm{d}}\left(\underline{\mathbf{h}}_{\mathrm{T},\mu_1,\mu_2}^{(5)}\right)$, the broadband noise reduction factor, $\xi_{\mathrm{n}}\left(\underline{\mathbf{h}}_{\mathrm{T},\mu_1,\mu_2}^{(5)}\right)$, and the broadband desired signal reduction factor, $\xi_{\mathrm{d}}\left(\underline{\mathbf{h}}_{\mathrm{T},\mu_1,\mu_2}^{(5)}\right)$, as a function of the broadband input SNR, for several values of $\mu_1 = \mu_2$. For a given broadband input SNR, the higher are the values of μ_1 and μ_2, the higher are the broadband gain in SNR and the broadband noise reduction factor, but at the expense of higher broadband desired signal distortion index and higher broadband desired signal reduction factor.

7.4.3 MVDR

The MVDR filter proposed by Capon [7], [8] is obtained by minimizing the MSEs of the residual interference-plus-noise subject to the distortionless constraints, i.e.,

$$\min_{\mathbf{h}_1^{(n)}} \left(\mathbf{h}_1^{(n)}\right)^H \mathbf{\Phi}_{\mathrm{in},2}^{(n-1)} \mathbf{h}_1^{(n)} \quad \text{subject to} \quad \left(\mathbf{h}_1^{(n)}\right)^H \boldsymbol{\rho} = 1, \tag{7.89}$$

$$\min_{\mathbf{h}_2^{(n)}} \left(\mathbf{h}_2^{(n)}\right)^H \mathbf{\Phi}_{\mathrm{in},1}^{(n)} \mathbf{h}_2^{(n)} \quad \text{subject to} \quad \left(\mathbf{h}_2^{(n)}\right)^H \mathbf{d} = 1, \tag{7.90}$$

where $\mathbf{\Phi}_{\mathrm{in},2}^{(n-1)}$ and $\mathbf{\Phi}_{\mathrm{in},1}^{(n)}$ are defined in (7.82) and (7.84), respectively. From the optimization of (7.89) and (7.90), we get

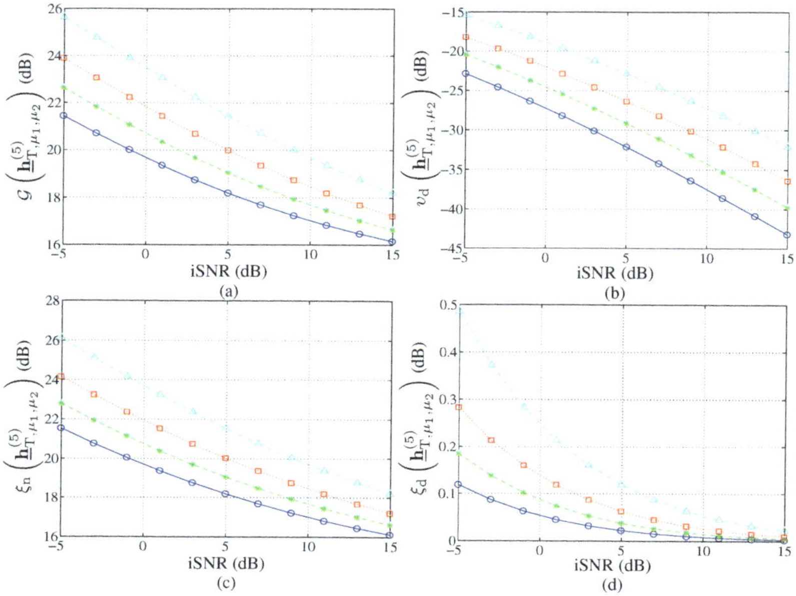

Fig. 7.4 (a) The broadband gain in SNR, (b) the broadband desired signal distortion index, (c) the broadband noise reduction factor, and (d) the broadband desired signal reduction factor of the tradeoff filter, $\underline{\mathbf{h}}^{(5)}_{\mathrm{T},\mu_1,\mu_2}$, as a function of the broadband input SNR, for $L = 2$, $M = 4$, and several values of $\mu_1 = \mu_2$: $\mu = 0.5$ (solid line with circles), $\mu = 1$ (dashed line with asterisks), $\mu = 2$ (dotted line with squares), and $\mu = 5$ (dash-dot line with triangles).

$$\mathbf{h}_1^{(n)} = \frac{\left(\mathbf{\Phi}_{\mathrm{in},2}^{(n-1)}\right)^{-1} \boldsymbol{\rho}}{\boldsymbol{\rho}^H \left(\mathbf{\Phi}_{\mathrm{in},2}^{(n-1)}\right)^{-1} \boldsymbol{\rho}} \tag{7.91}$$

and

$$\mathbf{h}_2^{(n)} = \frac{\left(\mathbf{\Phi}_{\mathrm{in},1}^{(n)}\right)^{-1} \mathbf{d}}{\mathbf{d}^H \left(\mathbf{\Phi}_{\mathrm{in},1}^{(n)}\right)^{-1} \mathbf{d}}, \tag{7.92}$$

with the initialization:

$$\mathbf{h}_2^{(0)} = \frac{\mathbf{\Phi}_{\mathbf{v}}^{-1}\mathbf{d}}{\mathbf{d}^H \mathbf{\Phi}_{\mathbf{v}}^{-1}\mathbf{d}}. \tag{7.93}$$

As a result, the MVDR filter is at iteration n:

$$\underline{\mathbf{h}}_{\mathrm{MVDR}}^{(n)} = \mathbf{h}_1^{(n)} \otimes \mathbf{h}_2^{(n)}. \tag{7.94}$$

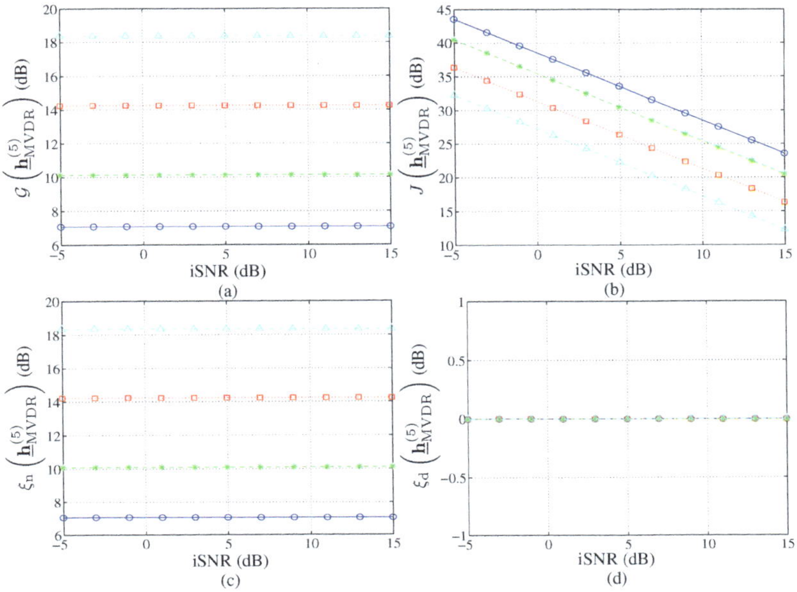

Fig. 7.5 (a) The broadband gain in SNR, (b) the broadband MSE, (c) the broadband noise reduction factor, and (d) the broadband desired signal reduction factor of the MVDR filter, $\underline{\mathbf{h}}_{\mathrm{MVDR}}^{(5)}$, as a function of the broadband input SNR, for $L = 2$ and different numbers of sensors, M: $M = 1$ (solid line with circles), $M = 2$ (dashed line with asterisks), $M = 4$ (dotted line with squares), and $M = 8$ (dash-dot line with triangles).

This filter can be directly obtained from $\underline{\mathbf{h}}_{\mathrm{T},\mu_1,\mu_2}^{(n)}$ by taking $\mu_1 = \mu_2 = 0$.

Example 7.3. Returning to Example 7.1, we now employ the MVDR filter at the iteration $n = 5$, $\underline{\mathbf{h}}_{\mathrm{MVDR}}^{(5)}$. Figure 7.5 shows plots of the broadband gain in SNR, $\mathcal{G}\left(\underline{\mathbf{h}}_{\mathrm{MVDR}}^{(5)}\right)$, the broadband MSE, $J\left(\underline{\mathbf{h}}_{\mathrm{MVDR}}^{(5)}\right)$, the broadband noise reduction factor, $\xi_{\mathrm{n}}\left(\underline{\mathbf{h}}_{\mathrm{MVDR}}^{(5)}\right)$, and the broadband desired signal reduction factor, $\xi_{\mathrm{d}}\left(\underline{\mathbf{h}}_{\mathrm{MVDR}}^{(5)}\right)$, as a function of the broadband input SNR, for $L = 2$ and different numbers of sensors, M. For a given broadband input SNR, as the number of sensors increases, the broadband gain in SNR and the broadband noise reduction factor increase, while the broadband MSE decreases. Compared with the Wiener filter, the MVDR filter yields no desired signal reduction, but at the expense of lower gain in SNR, lower noise reduction, and higher MSE (compare Figs. 7.2 and 7.5).

References

1. J. Benesty, J. Chen, and Y. Huang, *Microphone Array Signal Processing*. Berlin, Germany: Springer-Verlag, 2008.
2. J. Benesty, I. Cohen, and J. Chen, *Fundamentals of Signal Enhancement and Array Signal Processing*. Singapore: Wiley–IEEE Press, 2018.
3. M. Brandstein and D. B. Ward, Eds., *Microphone Arrays: Signal Processing Techniques and Applications*. Berlin, Germany: Springer-Verlag, 2001.
4. J. Benesty, J. Chen, and E. Habets, *Speech Enhancement in the STFT Domain*. Springer Briefs in Electrical and Computer Engineering, 2011.
5. Y. Avargel and I. Cohen, "On multiplicative transfer function approximation in the short-time Fourier transform domain," *IEEE Signal Process. Lett.*, vol. 14, pp. 337–340, May 2007.
6. J. Benesty, C. Paleologu, and S. Ciochina, "On the identification of bilinear forms with the Wiener filter," *IEEE Signal Process. Lett.*, vol. 24, pp. 653–657, May 2017.
7. J. Capon, "High resolution frequency-wavenumber spectrum analysis," *Proc. IEEE*, vol. 57, pp. 1408–1418, Aug. 1969.
8. R. T. Lacoss, "Data adaptive spectral analysis methods," *Geophysics*, vol. 36, pp. 661–675, Aug. 1971.

Index

J. Benesty et al., *Array Processing*, Springer Topics in Signal Processing 18, https://doi.org/10.1007/978-3-030-15600-8

The manufacturer's authorised representative in the EU is Springer Nature Customer Service Centre GmbH, Europaplatz 3, 69115 Heidelberg, Germany. If you have any concerns regarding our products, please contact ProductSafety@springernature.com

Printed and bound by CPI Group (UK) Ltd, Croydon, CR0 4YY
22/09/2026
02221063-0010